The Open University

Mathematics: A Second Level Course

An Algorithmic Approach to Computing   Units 6 and 7

ALGORITHMS AND THEIR CONSTRUCTION

COMPUTER HARDWARE

Prepared by the Course Team

THE OPEN UNIVERSITY PRESS

The Open University Press, Walton Hall, Bletchley, Buckinghamshire.

First published 1973.

Produced in Great Britain by
Technical Filmsetters Europe Limited, 76 Great Bridgewater Street, Manchester M1 5JY.

SBN 335 01292 2

This text forms part of the correspondence element of an Open University Second Level Course. The complete list of units in the course is given at the end of this text.

For general availability of supporting material referred to in this text, please write to the Director of Marketing, The Open University, Walton Hall, Bletchley, Buckinghamshire.

Further information on Open University courses may be obtained from The Admissions Office, The Open University, P.O. Box 48, Bletchley, Buckinghamshire.

1.1

036517

Unit 6   Algorithms and their Construction

Contents

Set Books

A. I. Forsythe, T. A. Keenan, E. I. Organick and W. Stenberg, *Computer Science: A First Course* (John Wiley, 1969).

A. I. Forsythe, T. A. Keenan, E. I. Organick and W. Stenberg, *Computer Science: BASIC Language Programming* (John Wiley, 1970).

It is essential to have these books to study this course. Throughout the correspondence texts, the set books are referred to as

> **A** for *Computer Science: A First Course*,
> **B** for *Computer Science: BASIC Language Programming*.

Notation

While we have used the notation of the set books for the representation of algorithms, we differ from them in the use of the symbols

> O for the capital letter O,
> Ø for zero.

This has been done to bring our texts into line with the notation you will use when writing programs. In addition, we sometimes use the symbol

> ∇ for a space

in a string of symbols.

6.0 INTRODUCTION

In this unit we shall complete the initial work of Units 1 to 4 on the construction of algorithms and their corresponding BASIC programs. The main aim of the unit is to help you to tackle larger problems. Many of the exercises in the unit cannot be completed without considerable effort on your part and you should not attempt them all unless you have sufficient time. Also, make sure that you check your answers to those that you do work on very carefully. An algorithm that is nearly right is nearly useless !

6.1 SUBROUTINES

6.1.1 Organizing an Algorithm

Human organizations set up to tackle a complex task almost invariably involve a hierarchy, with responsibility for specified tasks being passed down to subordinates.

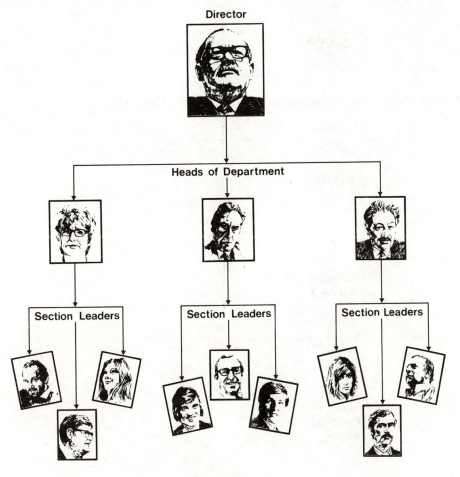

At each level of a managerial tree, the task in hand is split into sub-tasks and passed down to subordinates on the level below. Eventually, simple tasks are allocated at the lowest level and carried out. Those completed tasks are then marshalled at the next level up, to complete the next more complex tasks, and so on back up to the top of the tree.

The same sort of hierarchical organization can be used to construct computer programs. The hierarchy consists not of groups of people but of sections of program, each performing a specific task. These sections of program are usually called routines.

Just as the proper delegation of responsibility is a mark of good management, so the correct division of a program into routines is a mark of good programming. It is easy enough to give examples; for instance, many programs can be divided into an input routine, a calculation routine and an output routine.

This hierarchy is used in the construction of detailed flow charts. We could always begin with the same trivial flow chart.

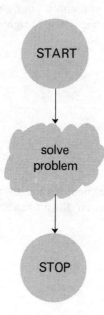

This can be broken down into several routines, giving a slightly more illuminating flow chart.

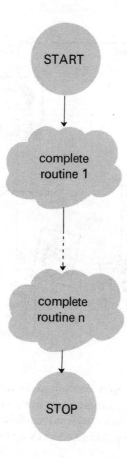

Each routine can then be broken down further until eventually a detailed flow chart is obtained and we have given our program a hierarchical structure similar to the management of a firm.

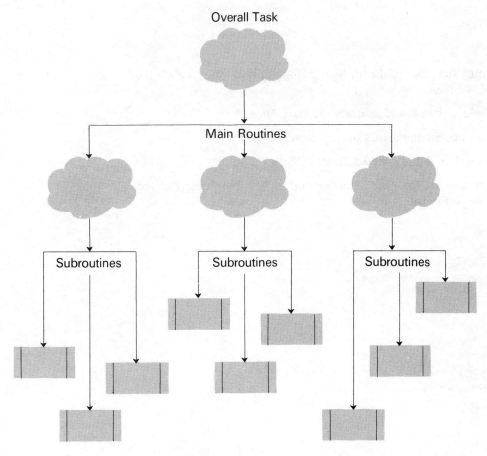

Overall Task

Main Routines

Subroutines   Subroutines   Subroutines

... and so on.

Working from the top down in this way means that we are thinking in terms of tasks less specific than the simple instructions of the final program. Yet we still write instructions, not now those instructions that the device could execute, but broader instructions such as DO THIS or TEST THAT.

We begin at the highest level with the imperious instruction

PERFORM TASK

As we work down through successive levels, to more and more detailed flow charts, the instructions become more precise and the demands less exacting. Thinking in this way about instructions at various levels of detail is a useful way of splitting a program up into routines. Each high-level instruction is a specification of what one routine must accomplish. If you find yourself lapsing into messy instructions like

FINISH THIS AND THEN DO PART OF THAT

it is likely that you have made an improper division into routines at that (or a higher) level.

Example

In a team sailing competition, six races are sailed and in each race, each boat is allocated points for the position in which it finishes, the most points going to the winner. To calculate the final score for a boat, only its five best races are considered; the worst score for each boat is ignored. A team consists of four boats but only the three highest scoring boats are counted in the final total for each team. The problem is to construct an algorithm which reads the points scored by each boat in a team and calculates the total score for a team.

Solution

Let us start by clarifying the problem. Each boat sails in six races and is awarded a score for each race. How then, can we identify the score that a particular boat gets in

7

a given race? The variable

A(I, J)

might be used to identify the score of the Ith boat in the Jth race. To calculate the Ith boat's overall score we then

sum A(I, J) for J = 1 to 6 and subtract the least A(I, J).

Call this sum B(I). To calculate the team's overall score we then

sum B(I) for I = 1 to 4 and subtract the least B(I).

Using this representation of the data, the following outline flow chart describes the routines that have to be performed.

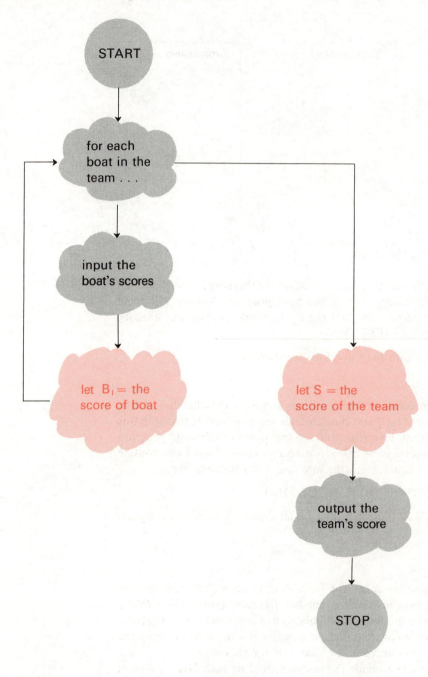

We shall leave this example for the moment and finish it later. It illustrates the point, commonly met in computing, that when you have managed to disentangle all the routines, certain routines may occur more than once, with no more than variations in the names of the variables they use. In a management hierarchy this situation corresponds to a number of different groups doing very similar tasks. They may all be needed if the total work-load is too great for any one individual to cope with. In this respect, the situation is radically different when we are constructing a program. For a routine can be used over and over again.

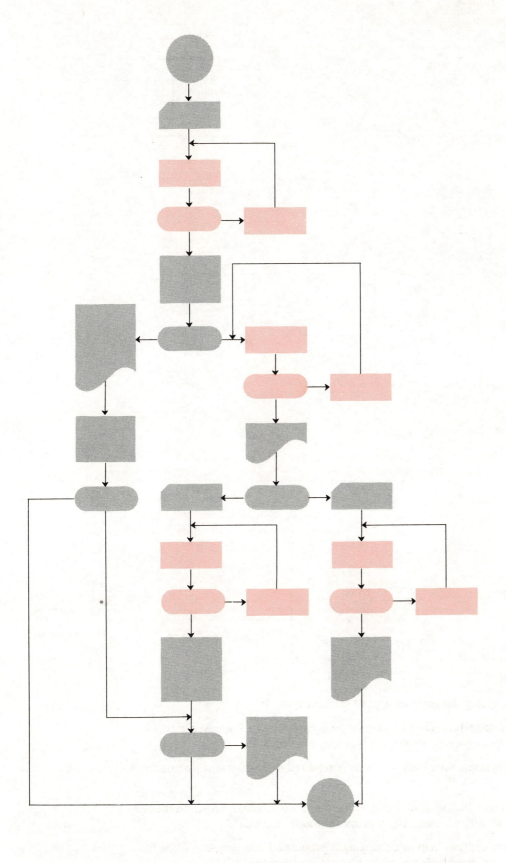

In the above algorithm there is a routine which is repeated four times. Clearly what we would like is a single copy of the routine and some way of digressing temporarily from the main stream of the algorithm. We would execute the routine and then return to the point at which we left the main stream. Routines which are used in algorithms in this way are called subroutines. This is how we denote them in a flow chart.

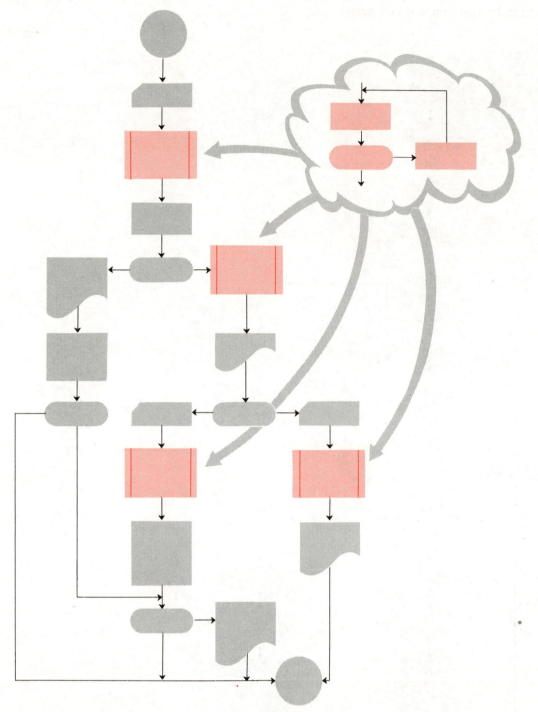

Some of the advantages of subroutines are readily understood:

(i)   They save time and effort. The instructions for a particular task are written once and for all and can then be used whenever they are required.

(ii) Storage space is saved since only one copy is required of instructions that occur many times.

(iii) They help to make the program easier to read and understand. Other routines are effectively shortened, especially if many subroutines are used.

(iv) Subroutines are of great help in debugging a program, since they can often be written and tested independently of the rest of the program.

(v)   A library can be built up of commonly used subroutines which are then fitted into any program that can make use of them.

Exercise 1

The following outline flow chart indicates the routines necessary to compute what notes and coins (£10, £5, £1, 50p, 10p, 5p, 2p, and 1p) make up a given amount.

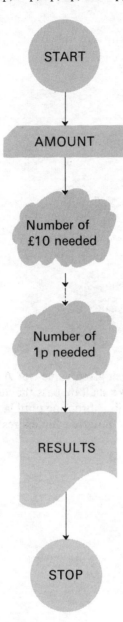

These routines have a common look about them. Draw a flow chart for a subroutine which will perform the required calculations*, using the following variable names

    A is the current amount of money,

    V is the value of note or coin in question,

    C is the number of coins of value V which are required.

---

* The calculations must find the "Number of ____ needed".

Solution 1

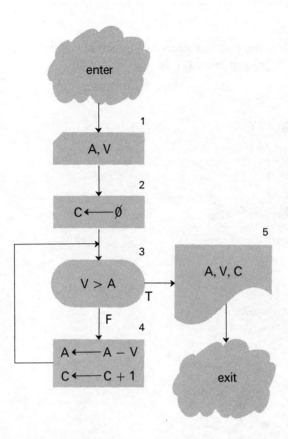

What we have constructed is a subroutine which, when given values for A and V as input, produces the values of A, V and C as output. We shall by-pass the question of how we digress from the main algorithm to execute this subroutine until later in this section. We shall also introduce a more satisfactory notation for expressing sub-routines.

The new feature of subroutines as distinct from other program instructions is that they can be shared by several parts of the program. We may get something like the following.

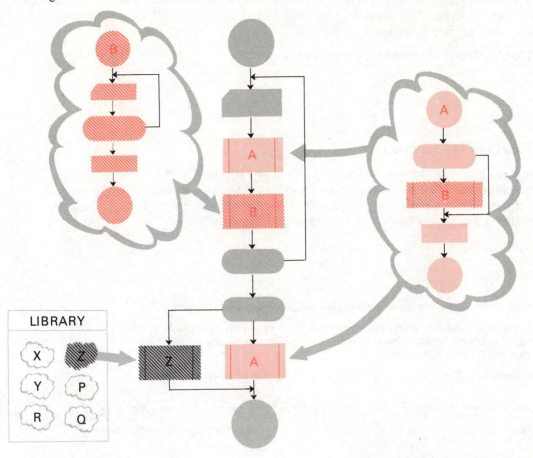

You can hardly fail to observe that any subroutine may be used at any level; for instance, the subroutine B is part of the main outline flow chart as well as being used by A. In this respect the use of subroutines is rather different from management hierarchy! What we shall guard against is the following.

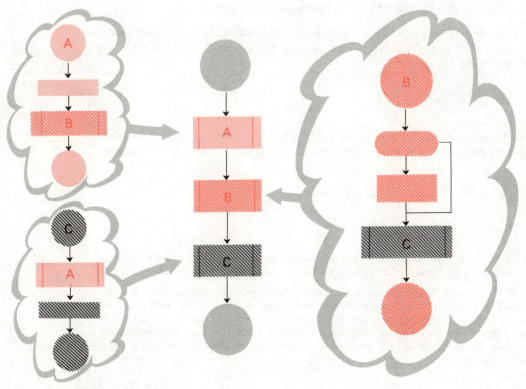

Here we have an algorithm which involves three subroutines, A, B and C. Consider A. Its execution is not complete before B is called in as a subroutine. Then B calls C, and C causes A to be restarted! So A will recur before its execution is complete—and this seems likely to happen yet again. It is possible to use recursive subroutines successfully, but great care is needed.

> *There's a hole in my bucket, dear Liza, dear Liza . . .*

6.1.2   Functions and Subroutines

We are now going to work through **A**, Chapter 6. It will give you a different slant on the idea of subroutines, but it is always good policy to look at things from more than one point of view. Since **A** was not primarily intended to be studied by mathematicians, some of you may find the explanation of mathematical ideas slightly lowbrow.

READ page A235 to page A238, line − 12

Notes

Page A235, line 1   *Reference flow charts* are flow charts of function routines or subroutines.

Page A236, Figures 6-3 and 6-5   These figures illustrate a source of possible misunderstanding arising from the notation adopted. What is to be fed into the funnel is not $\sqrt{y}$ but the value of y itself. The m which is returned is actually $\sqrt{y}$.

Page A236, Figure 6-5   The plaques inscribed U.S.1 and U.S.2 refer to American roads.

Exercise 2

Question 3 of Exercises 6-1, page A238.

Please do not read section 6-2, page A239 to page A246, line − 6. If you have any doubt about what functions are, we recommend you to read *An Introduction to Calculus and Algebra*, Volume 1 (Chapter 1) The Open University Press, 1971. If you need to, you should read Chapter 1 before going any further.

Most programming languages at a higher-level than machine language allow for a certain number of predefined functions (see page **B**17, Table 2-3). If we write the BASIC statement

```
10 LET Y=SIN (X)
```

its execution must inevitably involve the use of a built-in routine to compute SIN(X). The next reading passage describes how this built-in routine is executed by a computer.

READ page A246, line − 5, to page A253, line 17

Notes

Page A251, line 7   It is important that the name of the function be used to identify the subroutine. In the model, the assigner and reader will look for the brick chamber labelled (in this case) MIN.

Page A252, Figure 6-21   If you find this explanation difficult, ignore the B and C shown on the input funnel and, equally, the Z shown above the output window. The variables B, C and Z are supposed to be local to the subroutine and their names are not shown to the outside world. You might get a clearer idea of what actually happens if you imagine that there were two input funnels, one for (dare we say it!) B and the other for C. These could be labelled VARIABLE 1 and VARIABLE 2 respectively.

Exercise 3

Question 2 of Exercises 6-3, Set A, page A254*.

READ page A255, line −9, to page A262, line −8

Notes

Page A256, line 3   The word *procedure* is not restricted solely to routines for evaluating vector-valued functions.

Page A261, line 12   There is no good reason why the unlabeller should work outside the sealed chamber. It would be more in keeping with the spirit of the model if he removed the stickers before the boxes were dumped out of the chute.

Exercise 4

Exercises 6-4, Set C, page A264.

READ page A266, line −7, to page A271, line 6

Notes

Page A267, line 4   In this representation of complex numbers i represents $\sqrt{-1}$. A complex number is identified by two components both of which are real numbers. The first component stands on its own (as, for example, do a and c); the second component is always multipled by i (as, for example, b and d). Two complex numbers are equal only if their corresponding components are equal.

Exercise 5

Question 5 of Exercises 6-5 on page A272.

---

* A right triangle is a triangle in which one of the angles is 90°.

Solutions 2–5

2.   If $X \geq \emptyset$ then ABSOL(X) = X, whereas if $X < \emptyset$, ABSOL(X) = $-$X. The main
     feature of our flow chart must be a test of the sign of X. Little else is needed, and
     our reference flow chart might be as follows.

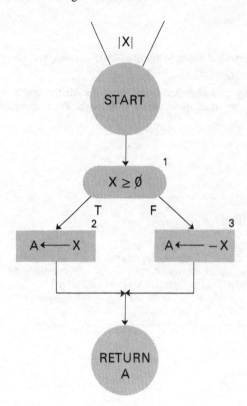

3. There are other ways of evaluating right (a, b, c), but the one given in the flow chart below is probably as good as any and better than most. We find the largest side and then check to see if it is the hypotenuse.

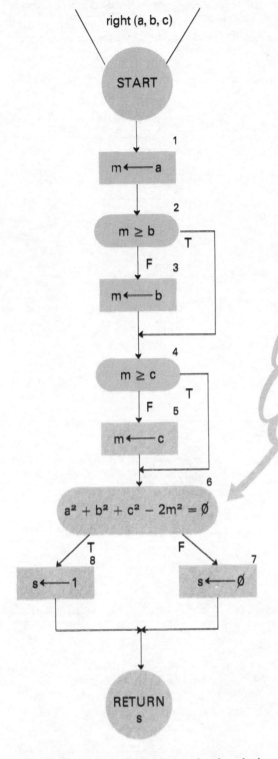

right (a, b, c)

START

1
$m \longleftarrow a$

2
$m \geq b$ T

F   3
$m \longleftarrow b$

4
$m \geq c$

F   5   T
$m \longleftarrow c$

At this stage, we do not know which of a, b or c is the greatest, only that m has the value of the greatest

6
$a^2 + b^2 + c^2 - 2m^2 = \emptyset$

T   8   F   7
$s \longleftarrow 1$ $s \longleftarrow \emptyset$

RETURN
s

4. There are a number of faults in the student's solution, some trivial, some less so.

(i) Box 2. The test for exit from this loop must be on $j < k$, not $j \leq k$. If the loop were iterated with $j = k$ it would refer to b_{k+1}, which does not exist.

(ii) Box 6. The exit must lead us back into box 3, not box 2. The inner loop deletes the second of two duplicates, i.e. b_{j+1}, and then "closes up the gap". If we then go to box 2 we shall increase j to $j + 1$ and so fail to check whether the new b_{j+1} is the same as b_j.

(iii) Box 6. Resetting k inside the outer loop is a very dangerous practice. According to the description of the iteration box given on page A140, it ought to

17

work in this instance, but, if you translated the flow chart into BASIC or some similar language, the result is unpredictable—it depends upon the compiler. Some compilers would set up the cut-off value on entry to the loop, using the value of k at that time; subsequent changes in k would not then influence the cut-off value. This is how BASIC on the *Student Computing Service* is compiled. The simplest way to avoid the danger we mention above is probably to recast the outer loop without using an iteration box. The corrected flow chart is as follows.

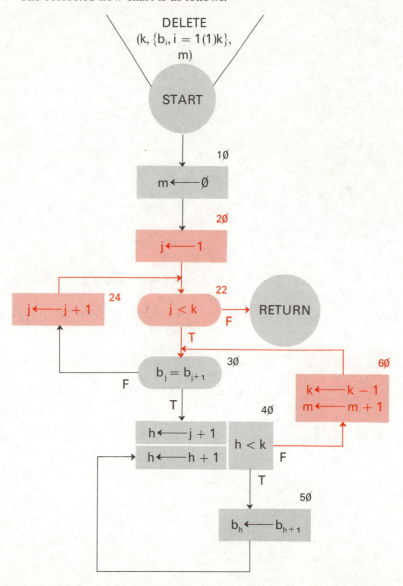

Whether you found this exercise hard or easy, it is worth reflecting for a moment how much more difficult it would have been to find the mistakes if this subroutine had been buried in a much larger flow chart. Similarly, by testing its constituent parts in isolation, one has a much better chance of successfully debugging a large program.

5. You should have had no difficulty in producing a flow chart like the following one.

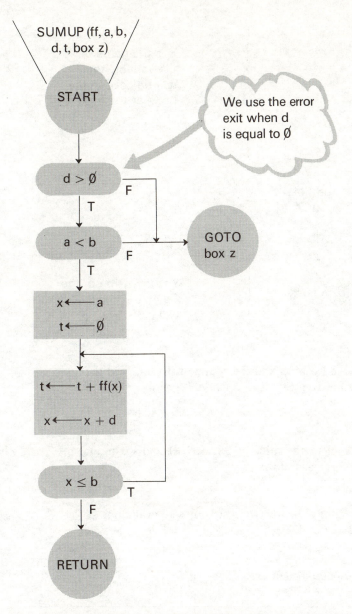

As you will have gathered from **A**, functions and subroutines have a number of points in common. They are both routines which are executed in isolation from the main body of the algorithm. They may both be executed at various stages of the algorithm. But most important, they both require parameters, that is, information which is passed to them by the main algorithm. However, they also differ in a number of ways, one being the type of parameter that we can pass to them. The reference flow charts as used in **A** emphasize how similar the two are, but we think it more important to emphasize their differences. This can best be done by modifying the notation for reference flow charts.

Suppose a function has been used in an arithmetic expression which is part of an assignment instruction. For example,

$$A \longleftarrow 4 * SIN(D + E) + B * C$$

In order to reflect the intended action of a routine to evaluate SIN(D + E), we draw the flow chart as follows.

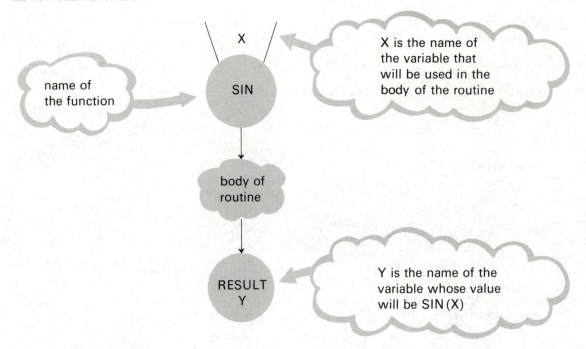

At the start of the function routine, the variable X in the routine will be passed the value of the expression on which the function SIN is to operate. If you like, think of this as an assignment.

$$X \longleftarrow D + E$$

When execution of the routine is complete, the value of Y is SIN(X), and so, in our example, we can replace SIN(D + E) by Y as follows.

$$A \longleftarrow 4 * Y + B * C$$

This process is explained in more detail by the following step-by-step evaluation of the expression, taking $D = E = \frac{1}{2}$, $B = 1$ and $C = 2$.

Step Number	Action	Appearance of Expression after Each Step	Remarks
1	Initial Expression	4 * SIN (D + E) + B * C	Identify first sub-expression
2	Compute D + E	4 * SIN (1) + B * C	Execute function routine
3	Compute SIN (1)	4 * .84 + B * C	
4	Compute 4 * .84	3.36 + B * C	

. . . and so on.

The important point to note about a function routine is that we pass a value to the routine and we have a value returned to us. The situation is different for subroutines.

Parameters are passed to the subroutine, these are processed and returned. In **A**, we have seen that we can pass

    (i)   values,

    (ii) variables,

    (iii) functions.

Thus we allow the following situation to occur.

List of parameters passed to the subroutine

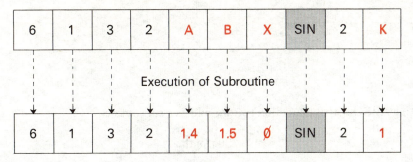

List of parameters returned by the subroutine

We recommend the following way of drawing flow charts for subroutines.

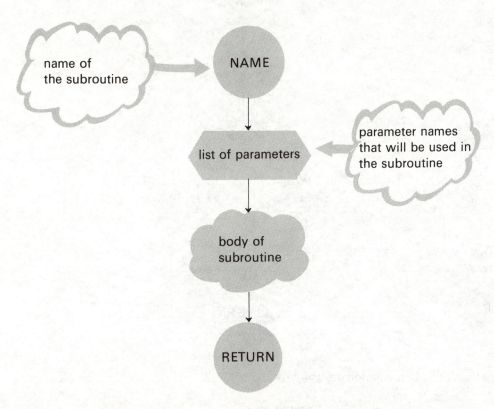

We shall use the same notation as **A** for calling subroutines in a flow chart.

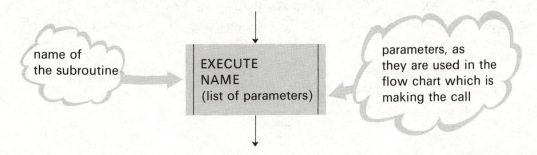

When using the written notation, we shall call subroutines as follows,

4.   execute NAME on (list of parameters)

and will head a description of both functions and subroutines as follows.

begin NAME on (list of parameters)

In place of the usual <u>end</u> statement for algorithms, we write

<u>return</u>

for subroutines and

<u>result</u> y

for function routines.

Example

Express a routine for evaluating ABSOL(X) as a function routine and as a subroutine.

Solution

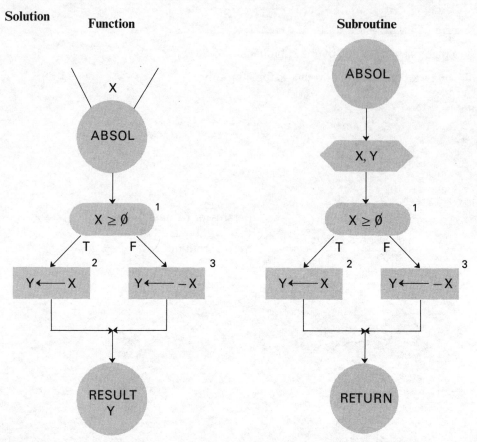

The corresponding routines in written notation are as follows.

Function	**Subroutine**
<u>begin</u> ABSOL <u>on</u> (X)	<u>begin</u> ABSOL <u>on</u> (X, Y)
1. <u>if</u> X ≥ ∅ <u>then</u> 2. Y ⟵ X	1. <u>if</u> X ≥ ∅ <u>then</u> 2. Y ⟵ X
<u>otherwise</u> 3. Y ⟵ − X	<u>otherwise</u> 3. Y ⟵ − X
<u>result</u> Y	<u>return</u>

Exercises 6–7

6. We now take another look at the example on page 7 of section 6.1.1. While dividing the algorithm into its principal routines we saw that two of them were very similar.

    (i)   Draw a flow chart for the scoring subroutines.

    (ii) Draw a detailed flow chart for the algorithm.

7. Question 5 of Exercises 6-4, Set A, page A263.

Solutions 6–7

6. (i) A neat way of working the flow chart for the subroutine is to add all the scores together and then subtract the smallest.

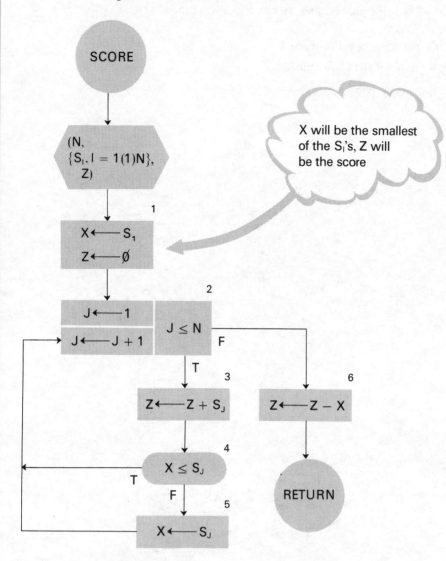

24

(ii) Using this subroutine, the detailed flow chart is

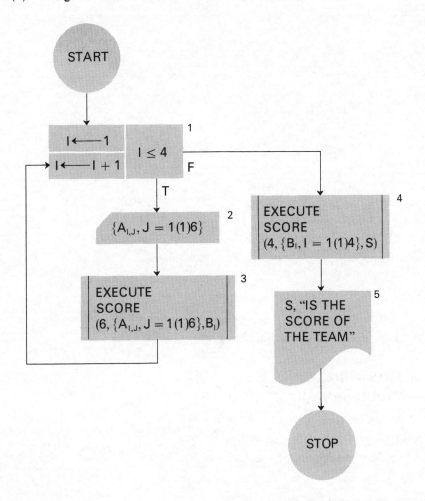

In this flow chart, we have used the notation given in the Example of section 6.1.1.

7. (a) Let us set up our answer to this part of the question as a subroutine called ALIQUO. The vector P* will be used to store the aliquot parts of a given number N. If we want the entries in P to be in ascending order, the following subroutine will suffice†.

the value of N will be passed to N1, A will record the number of factors of N1

```
begin ALIQUO on (N1, {P_J, J = 1(1)A}, A)
1.  P_1 ←—— 1 ; A ←—— 1
2.  for K = 2 by 1 to N1/2 ←————————————
        do  {3.  if N1 ≠ CHOP(N1/K) * K then ————
             4.  A ←—— A + 1 ; P_A ←—— K}
return
```

One problem might be that we do not know in advance what the dimension of P will be, that is, we do not know how many factors N will have. Fortunately, this does not cause any trouble when we come to write the corresponding BASIC program (see Exercise 9, section 6.1.3).

---

* P is the same as the vector PARTS referred to by A.

† Within the subroutine we should try to use different variable names from the main algorithm. This is often best done by adding a number to the name.

(b) Look back at the outline flow chart of Solution 12 of Unit 3, section 3.3.2. In that flow chart, we could replace

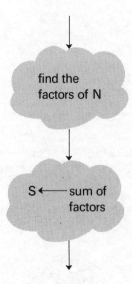

by a subroutine which is in turn called ALIQUO. Let us call this subroutine SUMPARTS.

begin SUMPARTS on (N2, S2)

1.   execute ALIQUO on (N2, $\{Q_J, J = 1(1)B\}$, B)

2.   S2 ⟵ Ø

3.   for L = 1 by 1 to B

     do {4.   S2 ⟵ S2 + Q_L}

return

(c) The algorithm can now be expressed as follows.

begin

1.   for N = 2 by 1 to 5ØØ ⟵

    do {2.   execute SUMPARTS on (N, S)

        3.   T ⟵ S

        4.   if T ≤ N then ⟶

        5.   execute SUMPARTS on (T, S)

        6.   if S ≠ N then ⟶

        7.   output N, "AND", T, "ARE FRIENDLY"}

end

Concerning line 4, if T = N, then N is perfect and hence not friendly (except with itself!). If T < N and T and N are friendly, the latter case will have been detected earlier (when N was what T is now!).

6.1.3 Functions and Subroutines in BASIC

As you already know, certain predefined functions are built into the BASIC language. Routines for evaluating these functions are provided as part of the *Student Computing Service*. Further details of functions and subroutines are given in **B**. Section 6-4 of **B** is not applicable to the *Student Computing Service* and so we ask you to omit it.

READ page **B**101 to page **B**107, line 6

Notes

Page B103, Figure 6-2   On the *Student Computing Service*, dummy lines are not allowed. There are two ways of amending the program. First, line 8Ø could be

```
80 GOSUB 1001
```

and second, a form of dummy line can be inserted at line 1ØØØ as follows.

```
1000 REM
```

Page B104, line 7   Replace "blank (dummy) line" by "REMARK statement".

Page B104, line 13   Any number of END statements can be used in a program and when used, they have the same effect as STOP statements. If, in addition, the highest numbered statement is not an END statement, the program will not run.

Page B104, lines 18–20   This is not strictly true of the *Student Computing Service*. Subroutines can be nested to a level of nine (see the *BASIC Reference Manual*); great care is needed to handle recursive subroutines and we advise against it.

Exercises 8–10

8. Write a BASIC program to compute what notes and coins are required to make up a given amount (see Exercise 1, section 6.1.1).

9. Write a BASIC program to compute a sailing team's score, according to the rules set out in the example of section 6.1.1 (see Exercise 6, section 6.1.2).

10. Question 5 of Exercises 6-4, Set A, page **B**112.

Solutions 8–10

8. Never overlook the facilities of the language you are using! It is not sensible to use a subroutine to express this algorithm as a BASIC program.

```
1    REM ***   NOTES AND COINS ALGORITHM   ***
5    INPUT A
7    FOR N=1 TO 8
10   READ V
20   LET C=0
30   IF V>A THEN 50
40   LET A=A-V
42   LET C=C+1
45   GOTO 30
50   PRINT C;"OF";V
55   NEXT N
80   DATA 10,5,1,.5,.1,.05,.02,.01
99   END
```

9. We have made the translations straight from the solution to Exercise 6 of section 6.1.2. First, the subroutine.

```
1000 REM   ***   SCORE SUBROUTINE   ***
1002 REM   ***   PARAMETERS N, (S(I), I=1(1)N), Z   ***
1005 DIM S(6)
1010 LET X=S(1)
1012 LET Z=0
1020 FOR J=1 TO N
1030 LET Z=Z + S(J)
1040 IF X<=S(J) THEN 1055
1050 LET X=S(J)
1055 NEXT J
1060 LET Z=Z-X
1065 RETURN
1070 REM   ***   SCORE ENDS HERE   ***
```

Notice that line 1002 is slightly different from the type of remarks made by **B**. You should always preface a subroutine with remarks of this kind.

Next, the main routine.

```
5 REM  ***  TEAM SCORE ALGORITHM  ***
7 DIM A(4,6), B(4)
10 FOR I=1 TO 4
20 REM  ***  INPUT BOAT'S SCORE  ***
22 FOR J=1 TO 6
24 INPUT A(I,J),
26 NEXT J
28 PRINT
30 REM  ***  EXECUTE SCORE ON 6, (A(I,J), J=1(1)6), B(I)  ***
31 LET N=6
32 FOR J=1 TO 6
33 LET S(J)=A(I,J)
34 NEXT J
35 GOSUB 1000
36 LET B(I)=Z
38 NEXT I
40 REM  ***  EXECUTE SCORE ON 4, (B(I), I=1(1)4), S  ***
41 LET N=4
42 FOR K=1 TO 4
43 LET S(K)=B(K)
44 NEXT K
45 GOSUB 1000
46 LET S=Z
50 PRINT S;"IS THE SCORE OF THE TEAM"
99 STOP
9999 END
```

In lines 3Ø and 4Ø, the remarks show which parameters from the main routine will be passed to the subroutine SCORE. Notice that for each of these parameters, there is a corresponding parameter given in line 1ØØ2 of SCORE.

10. (a) The following section of program is a direct translation of Solution 7(a), section 6.1.2.

```
1000  REM  ***  ALIQUO SUBROUTINE  ***
1002  REM  ***  PARAMETERS N1, (P(J), J=1(1)A), A  ***
1005  DIM P[100]
1010  LET P[1]=1
1012  LET A=1
1020  FOR K=2 TO N1/2
1030  IF N1#INT(N1/K)*K THEN 1045
1040  LET A=A+1
1042  LET P[A]=K
1045  NEXT K
1050  RETURN
1055  REM  ***  ALIQUO ENDS HERE  ***
```

(b) Before translating the flow chart of Solution 12, of Unit 3, section 3.3.2, we prepared a subroutine for SUMPARTS.

```
2000   REM  ***    SUMPARTS SUBROUTINE  ***
2002   REM  ***    PARAMETERS N2, S2  ***
2005   DIM Q[100]
2010   REM  ***    EXECUTE ALIQUO ON N2, (Q(J), J=1(1)B), B  ***
2012   LET N1=N2
2014   GOSUB 1000
2016   LET B=A
2017   FOR J=1 TO B
2018   LET Q[J]=P[J]
2019   NEXT J
2020   LET S2=0
2030   FOR L=1 TO B
2040   LET S2=S2+Q[L]
2045   NEXT L
2050   RETURN
2055   REM  ***    SUMPARTS ENDS HERE  ***
```

The BASIC program for finding perfect numbers which uses these subroutines is as follows.

```
5    REM  ***    PERFECT NUMBERS  ***
10   FOR N=2 TO 1000
20   REM  ***    EXECUTE SUMPARTS ON N, S  ***
22   LET N2=N
24   GOSUB 2000
26   LET S=S2
30   IF S#N THEN 45
40   PRINT N;"IS PERFECT"
45   NEXT N
99   END
```

If we now tried to run this program, an error would be detected since the highest numbered statement in our program is not END. We must add the following.

```
9999   END
```

(c) In order to find friendly numbers, we want to use the subroutines ALIQUO and SUMPARTS. We can most easily amend the perfect numbers program by typing

```
DEL - 1,99
```

followed by a new main routine.

```
5    REM  ***    FRIENDLY NUMBERS  ***
10   FOR N=2 TO 500
20   REM  ***    EXECUTE SUMPARTS ON N, S  ***
22   LET N2=N
24   GOSUB 2000
26   LET S=S2
30   LET T=S
40   IF T <= N THEN 75
50   REM  ***    EXECUTE SUMPARTS ON T, S  ***
52   LET N2=T
54   GOSUB 2000
56   LET S=S2
60   IF S#N THEN 75
70   PRINT N;"AND";T;"ARE FRIENDLY"
75   NEXT N
99   END
```

6.2 STRINGS

6.2.1 String Handling

When you see the word *computation* you probably think of arithmetic operations, with a few decisions thrown in for good measure; in other words, you think of numerical calculations. Yet, oddly enough, when we calculate with numbers, we work not with the values of the numbers but with the digits written on the page. For instance, to add two numbers together, we work digit by digit, usually from the least significant end. We can do this perfectly well without ever stepping back to observe the values of the entire numbers. What we do is to manipulate symbols. When we do algebra, these symbols include letters as well as digits and special symbols like $+$, $-$, $\times$, etc. Usually we give some meaning to a set of symbols arranged in a particular order. That is precisely what you are doing now: these symbols do have meaning, don't they!

A finite sequence of symbols is called a string and we can operate on strings in various ways. We can lengthen a string, shorten it, rearrange it and generally alter it in any way as may be useful.

If the thought has occurred to you that the operations one might wish to perform on strings may fittingly be entrusted to subroutines, then you should pat yourself on the back: this is indeed a very fruitful application of what we have been studying. It is in this light—as an application of subroutines—that you should view this section.

READ page A272, line -12, to page A277, line 19

Notes

Page A272, line -2   Decimal numbers are so familiar to us that we tend to forget that this is just a way of representing a number, and is not to be confused with the number itself. We speak of the number 229, but this is really a character string and the same number could equally well be represented by the (binary) character string 11100101.

Exercise 11

Question 2 of Exercises 6-6, page A277.

Solution 11

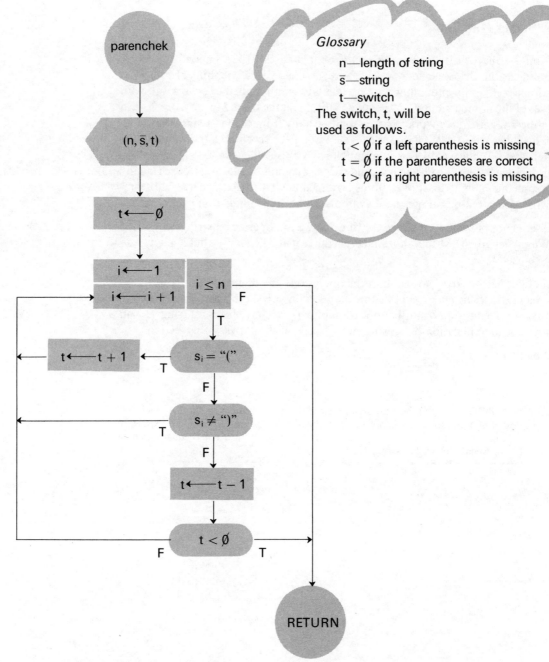

Glossary

n—length of string
$\bar{s}$—string
t—switch

The switch, t, will be used as follows.

$t < \emptyset$ if a left parenthesis is missing
$t = \emptyset$ if the parentheses are correct
$t > \emptyset$ if a right parenthesis is missing

6.2.2 Strings in BASIC

The string facilities described in **B**, section 6-6 (pages **B**115–116) are not available on the *Student Computing Service* and so we suggest you omit that section. We shall therefore describe some of the string processing operations that you can use and which are sufficient to allow you to tackle a variety of non-numerical problems.

In Unit 2, section 2.1.2, we saw that a string of characters can be assigned to a variable and we append $ to the name of a string variable. Every string variable in a program must be declared in a DIMENSION statement if it is to be assigned more than one character. For example, the statement

```
10 DIM A $(50), B $(72)
```

would declare that the variables A$ and B$ can be used to store up to 50 and 72 characters respectively.

As with arrays, the dimension must be given as a constant. DIM A$(N) is not allowed. The dimension of a string indicates the maximum number of characters it may contain. It is not necessary to use all the locations reserved; the string will be filled, however, from left to right. The maximum size of any string is 72 characters.

A string variable can be compared with another variable of the same type or with a string constant. But remember that when comparing two strings, the first operand may not be a constant. That is,

```
10 IF "CAT"=A$ THEN 100
```

is illegal, whereas

```
10 IF A$="CAT" THEN 100
```

is allowed. Also, compound assertions may not be built up from simple assertions involving string operands. We now want to consider what else we can do with strings on the *Student Computing Service*. For example, given the sentence

"CATCH THE BROWN FOX THAT JUMPED OVER THE DOG",

how do we find the number of words in it? Does it contain the word DOG? The following facilities help us to answer these questions.

(i) Substrings

A substring identifies part of a string. We write A$(I, J) to mean those characters from the Ith up to and including the Jth in the string A$. Thus

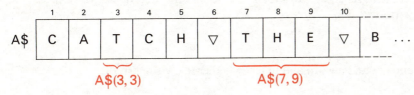

If J = I, A$(I, I) identifies the Ith character in the string. If J is omitted, the substring runs from I to the end of the string. We can use the substrings as follows. The statement

```
100 LET B$=A$(I,J)
```

will assign the substring A$(I, J), to the string B$ while

```
100 LET A$(I,J)=B$
```

will assign the string B$ to the substring A$(I, J). In this case if I − J + 1 is greater than the number of characters in B$ then the substring A$(I, J) would be filled on the right with the appropriate number of spaces. If on the other hand I − J + 1 is less than the number of characters in B$ then only the first I − J + 1 character of B$ will be inserted into the substring of A$.

(ii) Null String

The null string is a string which has no characters. There are various ways of testing for or assigning the null string. Perhaps the simplest is to use "" to identify the null string. Thus the assertion

```
100 IF A$ = "" THEN 200
```

will enable us to check if A$ is the null string, while

```
100 LET A$=""
```

will assign the null string to A$.

(iii) **Length of a String**

The DIMENSION statement identifies the maximum number of symbols allowed in any string. To check for the number of symbols in a string at any time the LEN function is used.

    LEN (A$)

is an integer between Ø and 72, the length of the null string being zero.

(iv) **Data**

When a string is included in a DATA statement, the symbols must be included within quotation marks.

```
10 READ A, B$, C$, D
200 DATA 15.4,"ABCDEFGH","1.2345",1.2345
```

If a simple INPUT statement is used, for example

```
10 INPUT A$
```

the symbols typed at the teletype are taken as the symbols of the string. When, however, input is mixed, for example

```
10 INPUT A, B$, C$, D
```

we have to type the input exactly as in the DATA statement, including the quotation marks.

Let us have a look at some of these ideas in an example. Suppose we want to write a program to search for a word in a sentence. This sort of routine would be quite fundamental to an interactive teaching program, that is a program in which the student would be allowed to type in his answer to a question—or his own question—which would then be analysed for content. As a key word is spotted so an interpretation of the student's response can be made.

Example

Write a program to read a sentence and a word and to test whether the word is in the sentence.

Solution

We have first to interpret the question. The sentence and word are both strings, so essentially we have to test whether one string is a substring of another. Given two strings A$ and B$, B$ will be a substring of A$ whenever the following assertion is true, for some suitable values of I and J.

$$B\$ = A\$(I, J)$$

Using this assertion, the core of the algorithm will be as follows.

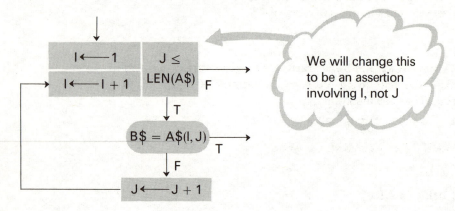

We will change this to be an assertion involving I, not J

It is now a fairly simple matter to expand this flow chart fragment to a detailed flow chart for the whole algorithm.

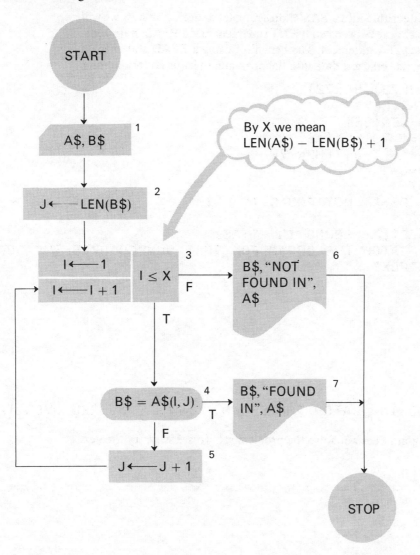

To test this algorithm, let us trace it for a simple case, say when A\$ and B\$ are both the same, which is a bit extreme but a good test to make.

Step Number	Instruction Number	Link Field	Values of Variables							Decisions				
			A\$	B\$	J	I				I ≤ X	B\$ = A\$(I, J)			
1	1	input 1, 2	CAT	CAT										
2	2				3									
3	3					1				T				
4	4										T			
5	7	output 1												
		STOP found												

This test does not prove that the algorithm is correct, but since all the assertions were tested and all the variables were used, it is worth proceeding to the BASIC program.

Before we translate the algorithm into a BASIC program, let us decide on how we will deal with box 1. We can choose between an INPUT statement and a READ statement; the latter will require a DATA statement. The benefit of using a READ statement is that we will be able to build some test data into the program, so let us go for that one.

```
10   DIM A$[72],B$[72]
12   READ A$,B$
20   LET J=LEN(B$)
30   FOR I=1 TO LEN(A$)-J+1
40   IF B$=A$[I,J] THEN 70
50   LET J=J+1
55   NEXT I
60   PRINT B$;"  NOT FOUND IN  ";A$
65   STOP
70   PRINT B$;"  FOUND IN  ";A$
80   DATA "CATCH THE BROWN FOX THAT JUMPED OVER THE DOG"
82   DATA "FOX"
99   END


     RUN


     FOX  FOUND IN  CATCH THE BROWN FOX THAT JUMPED OVER THE DOG
```

Unfortunately this program does not solve the problem set. To see why, try the next exercise.

Exercise 12

Modify the program opposite to test whether the words FOX, DOG, RAT and CAT are substrings of the test string A$. Choose a suitable terminating condition.

Solution 12

Looking at the flow chart on page 35, we see that it is the input box that needs modification.

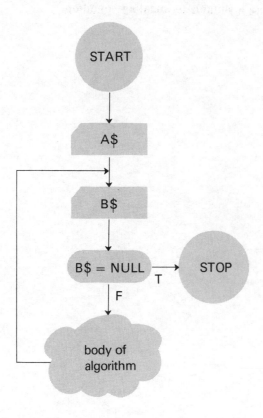

The appropriate BASIC statements are as follows

```
12 READ A $
14 READ B $
16 IF B $="" THEN 99
```

We also need to make the following amendments to the rest of the program

```
65 GOTO 14
75 GOTO 14
82 DATA "DOG"
83 DATA "OVER"
84 DATA "RAT"
85 DATA "CAT"
86 DATA ""
```

When we come to run the amended program, one line of output is

```
CAT  FOUND IN  CATCH THE BROWN FOX THAT JUMPED OVER THE DOG
```

Clearly our program has a bug in it. We are only testing whether B$ is a substring of A$, not whether B$ is a word of the sentence A$. When B$ is CAT, we really want to test if

$\triangledown$CAT$\triangledown$

is a substring of A$; that is, we need to add a space to both ends of B$. But this will not work if B$ is the first or last word of A$.

We can cope with the latter problem by adding a space to both ends of A$. Since both operations are the same, we would do well to think of performing this task by means of a subroutine.

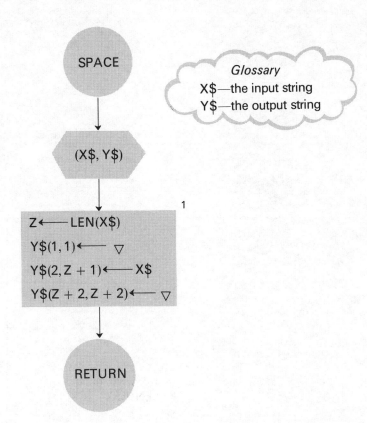

What does the flow chart look like now?

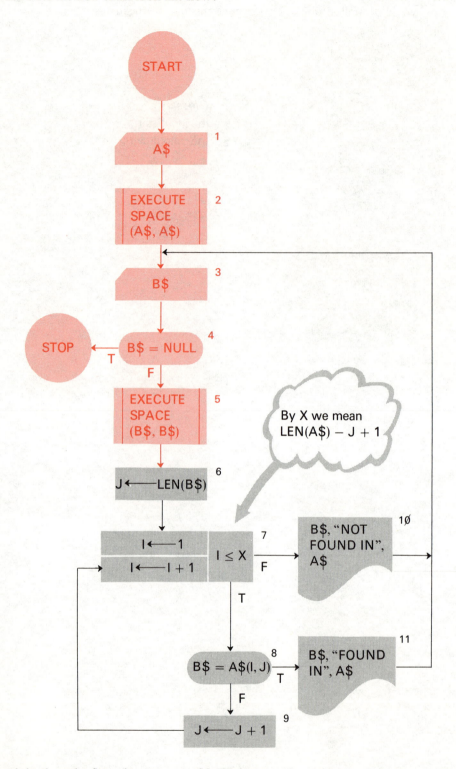

Look back at the flow chart on page 35. We have made a number of changes to the algorithm in the light of tests and bugs that we found. If we had made the alterations to the program only, it would have been easy to lose sight of the overall scheme of the algorithm. By amending the flow chart, those parts of the algorithm that are unchanged stand out.

All that we now need to do is to express this new flow chart as a BASIC program.

```
5    DIM A$[72],B$[72],X$[72],Y$[72]
10   READ A$
20   REM   ***  EXECUTE SPACE ON A$, A$   ***
23   LET X$=A$
24   GOSUB 1000
26   LET A$=Y$
30   READ B$
38   REM   ***  TEST FOR LAST B$   ***
40   IF B$="" THEN 9999
50   REM   ***  EXECUTE SPACE ON B$, B$   ***
53   LET X$=B$
54   GOSUB 1000
56   LET B$=Y$
60   LET J=LEN(B$)
68   REM   ***  LOOP STARTS HERE   ***
70   FOR I=1 TO LEN(A$)-J+1
80   IF B$=A$[I,J] THEN 110
90   LET J=J+1
95   NEXT I
100  PRINT B$;"  NOT FOUND IN   ";A$
105  GOTO 30
110  PRINT B$;"  FOUND IN   ";A$
115  GOTO 30
200  DATA "CATCH THE BROWN FOX THAT JUMPED OVER THE DOG "
202  DATA "FOX"
204  DATA "OVER"
206  DATA "RAT"
208  DATA "CAT"
210  DATA ""
1000 REM   ***  SPACE SUBROUTINE   ***
1002 REM   ***  PARAMETERS X$, Y$   ***
1010 LET Z=LEN(X$)
1015 LET Y$=""
1020 LET Y$[1,1]=" "
1030 LET Y$[2,Z+1]=X$
1040 LET Y$[Z+2,Z+2]=" "
1050 RETURN
1055 REM   ***  SPACE ENDS HERE   ***
9999 END
```

Exercise 13

Write a BASIC program to search a sentence for given nouns, insert a specified adjective before each noun which has been located and print the modified sentence.

Solution 13

We start with an outline flow chart.

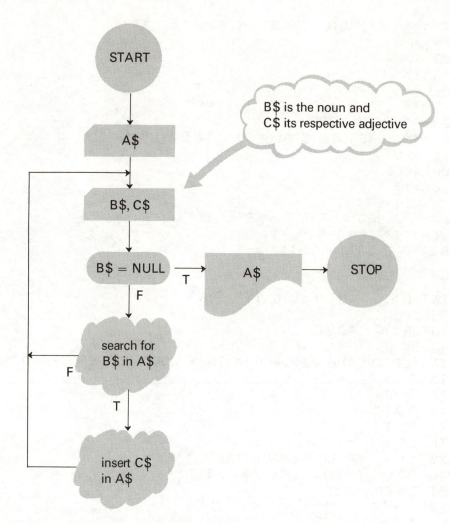

Most of this algorithm we have seen before. All we have to do that is new is to insert C\$ in A\$. This can be done as follows,

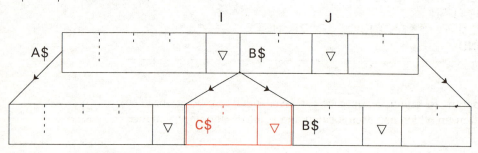

or, more explicitly, by the following assignments.

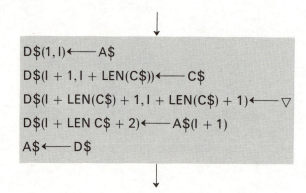

D\$(1, I) ⟵ A\$

D\$(I + 1, I + LEN(C\$)) ⟵ C\$

D\$(I + LEN(C\$) + 1, I + LEN(C\$) + 1) ⟵ ▽

D\$(I + LEN C\$ + 2) ⟵ A\$(I + 1)

A\$ ⟵ D\$

If we replace the PRINT statements in the program on page 41 by these assignments, and make the necessary alterations to the READ and DATA statements, we can amend the old program to give us the following one.

```
5    DIM A$[72],B$[72],X$[72],Y$[72]
7    DIM C$[72],D$[72]
10   READ A$
20   REM  ***  EXECUTE SPACE ON A$, A$  ***
23   LET X$=A$
24   GOSUB 1000
26   LET A$=Y$
30   READ B$,C$
38   REM  ***  TEST FOR LAST B$  ***
40   IF B$="" THEN 9000
50   REM  ***  EXECUTE SPACE ON B$, B$  ***
53   LET X$=B$
54   GOSUB 1000
56   LET B$=Y$
60   LET J=LEN (B$)
68   REM  ***  LOOP STARTS HERE  ***
70   FOR I=1 TO LEN (A$)-J+1
80   IF B$=A$[I,J] THEN 110
90   LET J=J+1
95   NEXT I
97   GOTO 30
110  REM  ***  INSERT C$ IN A$  ***
112  LET X=LEN (C$)
120  LET D$[1,I]=A$
130  LET D$[I+1,I+X]=C$
140  LET D$[I+X+1,I+X+1]=" "
150  LET D$[I+X+2]=A$[I+1]
160  LET A$=D$
165  GOTO 30
200  DATA "CATCH THE FOX THAT JUMPED OVER THE DOG"
202  DATA "FOX","BROWN"
204  DATA "OVER","HIGH"
206  DATA "RAT","GREY"
208  DATA "CAT","BLACK"
210  DATA "",""
1000 REM  ***  SPACE SUBROUTINE  ***
1002 REM  ***  PARAMETERS X$, Y$  ***
1010 LET Z=LEN (X$)
1015 LET Y$=""
1020 LET Y$[1,1]=" "
1030 LET Y$[2,Z+1]=X$
1040 LET Y$[Z+2,Z+2]=" "
1050 RETURN
1055 REM  ***  SPACE ENDS HERE  ***
9000 PRINT A$
9999 END
```

6.3 PROBLEM SOLVING

6.3.1 What is an Algorithm?

In the first four units of this course, we have studied the fundamental instructions used to construct algorithms; how to translate these instructions into the BASIC programming language; in general, how to solve some fairly straightforward kinds of problem. In this section we want to step back and take a look at where our work so far has brought us—to reflect on what we have seen.

Suppose we were given the following problem.

Find the roots of $x^2 + 3x + 2 = 0$.

Here are two possible solutions.

Solution 1

$$x^2 + 3x + 2 = (x + 1)(x + 2)$$

$$= 0 \; when \; x = -1 \; or \; -2.$$

The roots are -1 and -2.

Solution 2

1. *Set $a = 1, b = 3, c = 2$.*
2. *If $b^2 - 4ac < 0$ then say so and stop.*
3. *Set $x_1 = \dfrac{-b + \sqrt{b^2 - 4ac}}{2a}, x_2 = \dfrac{-b - \sqrt{b^2 - 4ac}}{2a}$.*
4. *The roots are x_1, x_2.*
5. *Stop.*

Solution 1 shows the direct method. The problem asks for two real numbers (if they exist) and the last line states what they are. Solution 2, on the other hand, shows the constructive method. If the steps outlined there were to be carried out, the required two numbers would be found (if they exist). Our interest lies in the constructive method, for it leads us to algorithms. When we talk about algorithms, we mean more than is immediately apparent from the formal definition given in Unit 1, which is as follows.

> An algorithm for the solution of a problem by a device is a specification of a finite number of instructions such that if the instructions are executed by the device then either a solution of the problem is found, or it is determined that no solution exists. In either case, execution stops after a finite number of steps.

For an algorithm to be of any use, we also want the following to be true.

(i) An algorithm has one or more inputs, that is, quantities which are given to it before it is executed. These inputs are restricted both by the conditions of the problem and the nature of the device. Part of an algorithm must determine if the input is acceptable.

(ii) An algorithm produces one or more outputs, that is, quantities which have a specified relationship to the inputs. The outputs are the result of executing the algorithm; this is why every algorithm must have at least one output. Performing an algorithm which led to no output would clearly be a meaningless activity.

(iii) The condition of finiteness is not really strong enough to be of practical use. A useful algorithm must not only be finite, it must be *very* finite, that is, its execution must end in a reasonable period of time. There are many algorithms which satisfy all the formal conditions but which we know will never be executed in our lifetimes even though the problems they solve are of interest to many people, simply because of the fantastically large number of steps they require to be executed.

(iv) An algorithm will require the device to use storage. The storage capabilities of the device form an important restriction on the usefulness of an algorithm.

(v) Above all, we require an algorithm to be efficient and adaptable. In Unit 3, section 3.3.3, we began to set up some formal criteria for comparing the efficiency of equivalent algorithms; adaptability is harder to describe formally. Informally, an algorithm only capable of adding 1, 2 and 3 would be useless; one capable of adding a, b and c might have its uses.

These ideas are always worth bearing in mind when you are constructing algorithms.

6.3.2 Methods of Problem Solving

Any detailed description of how to solve problems would inevitably read as an account of the author's personal fancy. But in broad outline there are a number of points that people can agree upon. Already in the course we have given quite a few items of specific advice. In the solutions to the exercises, we have, where appropriate, shown how the advice can be put into practice. In turn, you should by now have found which items you find useful and you should be adopting a more global attitude to solving a specific problem than the problem itself would appear to warrant. What we would like to do in this section is to reinforce that attitude with a few new remarks and a summary of what you have seen. A word of warning however about what we shall say! The processes that we recommend may be carried out in a matter of seconds, they may take minutes or even hours. This timing will certainly vary immensely from problem to problem. Also, the remarks are not guaranteed to be appropriate in every situation. In the end, it is up to you to develop your own strategies—we can only outline some advice that others have found useful.

Problems however phrased are seldom couched in terms that are immediately clear. Before we can start thinking about a solution we must be sure that we understand what the problem is about. Also, we ought to make sure that the problem is what we think it is. This can be helped if we are prepared to analyse the problem into three parts

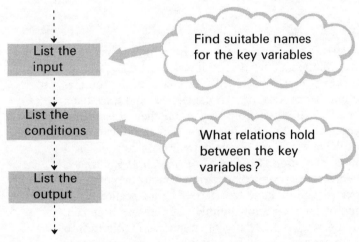

Having completed an initial analysis of the problem, ask yourself

"Can I see how to solve it?"

When the answer is "No", there is no need to abandon hope or to sit waiting hopefully for inspiration. There are some procedures that are worth following. First, try some simple cases of the problem: even try some extreme cases and put in some particular values. When doing any one of these be on the look out for a pattern that repeats itself. The results of a particular case may suggest a process that could be worked up into a method of solution. If this fails, try to think whether your problem is related to one you have solved previously. The method of solution to a similar problem may indicate a method of solution to the current problem. If this does not help, it may be time to go back and reanalyse the problem to make sure that you really understand it. When the answer is "Yes", go on to produce an outline flow chart or to develop the core of the algorithm (see Unit 3, section 3.0).

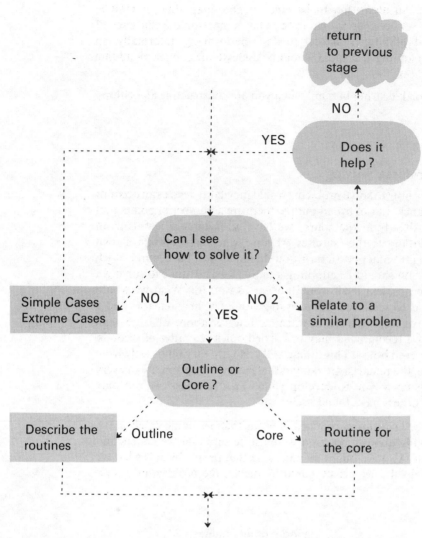

We have been discussing how to devise a plan for the solution. But however successful you are at thinking up a method of solution all may be lost if you do not carry out the plan with care and thoroughness. First prepare detailed flow charts of the routines that make up the complete solution. Be sceptical of their correctness and trace them with test data. If the test data is to be effective you should make sure that all assertions are tested and all variables assigned values. It is often hard to specify the exact criteria of the assertions without tracing. We have already given advice on how to translate a flow chart into a BASIC program (see Solution 11 of Unit 2, section 2.1.3). But what if the program will not run? Simple errors in a program are pointed out by the computer system; these can usually be corrected without much trouble (see Unit 2, section 2.2.2). However, bugs in a program do not show up until suitable test data has been run. Debugging is not so easy (see Unit 4, section 4.3.1). It is often important to return to the detailed flow charts to make the necessary amendments. What at first may appear to be a simple amendment may have a serious knock-on effect which is hard to spot if the overall scheme of the algorithm has been obscured by the bleak format of a BASIC program.

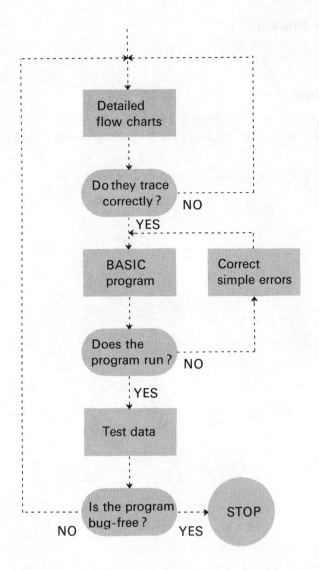

A summary of this section is given at the back of this unit by the flow chart given there. We would not suggest that the flow chart represents an algorithm. There is no way in which it represents *a specification of a finite number of instructions such that if the instructions are executed . . .* What is missing is the element of judgement. This you will acquire with time. Also missing is this button.

We leave it to you to decide on the appropriate moment to use it.

6.3.3 The Monotone Sequence Problem

READ NO MORE THAN page A191, line −6, to page A192, line 3, and CLOSE THE BOOK

Let us repeat the statement of the problem.*

> *Problem.* Suppose you are given a sequence (that is, a list) of numbers, all guaranteed to be different. Prove that the length of the longest monotone subsequence is at least $\sqrt{N}$.

Since this course is not concerned with proving results, but with finding results, we ask you to tackle a related sub-problem.

* A mathematical proof of this problem is given in *Appendix I*.

Exercise 14

For a given sequence, find the length of its longest monotone subsequence.

Notes

Sequence: a collection of N numbers listed in a particular order.

Subsequence: a subset of the original sequence, listed in the same order as in the original. Note that a subsequence is itself a sequence.

Length of a sequence: the number of entries in the list. The original sequence is of length N.

Monotone sequence: one in which either every entry apart from the first is greater than the one immediately before or every entry apart from the first is less than the one immediately before it.

We hope that you will allow yourself plenty of time to think about this problem. The advice in the previous section should help.

Solution 14

READ page A192, line 4, to page A198, line 7

Notes

Page A192, line −4   The rest of the reading passage is devoted to solving the problem as stated here. In order to complete Exercise 14, we will have to go a little further, but note that the problem that we are about to discuss is just a slightly simpler one than that posed in the exercise. It is often a useful strategy, if a problem looks complicated, to find and solve a simpler problem first.

Page A193, line 7   It is quite common in problems of this kind to find that the brute force approach leads to a computation of utterly impractical length.

Page A193, line 14   Note that it is I which goes from 1 to 9, and that we are looking for the longest increasing subsequence ending in each A_I. The idea for this approach comes from thinking about simple cases. We start with short sequences, look for a pattern and extend it to longer sequences.

Page A194, line 6   Here it is natural to start working on the core of the algorithm. The adage is:

"Start in the middle, go on with the end and end with the beginning".

Page A195, Figure 4-50   Presumably, if you use our terminology, you would start Figure 4-50 as follows,

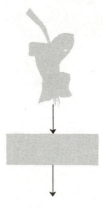

and add the sub-title:

Figure 4-50 Core of the Algorithm.

Page A197, line 8   It is wise never to be content with the algorithm as it is first obtained. When devising an algorithm you have the opportunity, as it were, to be wise after the event and to improve upon your first thoughts.

In the flow chart of Figure 4-53, page A199, we see the solution to half of our problem. The following flow chart completes the solution in a very natural way.

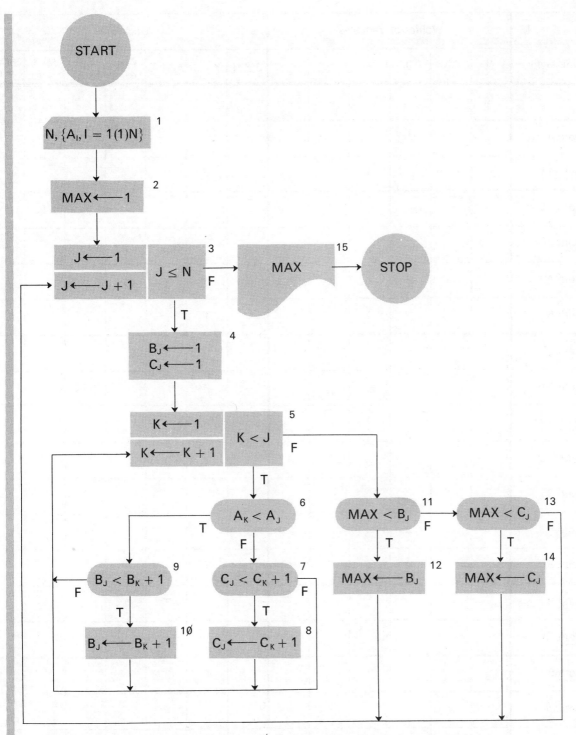

If we choose as test data the sequence Ø, 2, 1 for which N = 3 the trace table is not unmanageably long, but we do test all the assertions.

Step Number	Instruction Number	Link Field	Values of Variables						Decisions						
			N	MAX	J	K			J ≤ N	K < J	$A_K < A_J$	$C_J < C_K + 1$	$B_J < B_K + 1$	MAX < B_J	MAX < C_J
1	2		3	1											
2	3	goto 4			1				T						
3	5	goto 11				1				F					
4	11	goto 13												F	
5	13	goto 3													F
6	3	goto 4			2				T						
7	5	goto 6				1				T					
8	6	goto 9									T				
9	9	goto 1∅											T		
1∅	5	goto 11				2				F					
11	11	goto 12												T	
14	12	goto 3		2											
13	3	goto 4			3				T						
14	5	goto 6				1				T					
15	6	goto 9									T				
16	9	goto 1∅											T		
17	5	goto 5				2				T					
18	6	goto 7									F				
19	7	goto 8										T			
2∅	5	goto 11				3				F					
21	11	goto 13												F	
22	13	goto 3													F
23	3	goto 15			4										
24	15	output 1		2											
		STOP found													

The corresponding input, output and variables lists are as follows.

Input List		
Link	Value	Variable
1	~~3~~	N
2	~~Ø~~	A_1
3	~~2~~	A_2
4	~~1~~	A_3

Output List		
Link	Variable	Value
1	MAX	2

Variable	Current Value							
A_1	Ø							
A_2	2							
A_3	1							
B_1	1							
B_2	~~1~~	2						
B_3	~~1~~	2						
C_1	1							
C_2	1							
C_3	~~1~~	2						

Having verified the flow chart by tracing, we can now go ahead to write it as a
BASIC program.

```
5 REM  ***   MONOTONE SEQUENCE PROBLEM  ***
7 DIM A(100), B(100), C(100)
10 PRINT "PLEASE INPUT THE LENGTH OF THE SEQUENCE"
11 INPUT N
13 PRINT "PLEASE INPUT THE SEQUENCE"
14 FOR I=1 TO N
15 INPUT A(I)
16 NEXT I
20 LET M=1
30 REM  ***  MAJOR LOOP STARTS HERE  ***
32 FOR J=1 TO N
40 LET B(J)=1
42 LET C(J)=1
50 REM  ***  LOOP TO FIND LONGEST SO FAR ***
52 FOR K=1 TO J-1
60 IF A(K) < A(J) THEN 90
70 IF C(J)>=C(K)+1 THEN 105
80 LET C(J) =C(K)+1
85 GOTO 105
90 IF B(J)>=B(K)+1 THEN 105
100 LET B(J)=B(K)+1
105 NEXT K
107 REM  ***   LONGEST SO FAR ENDS HERE  ***
110 REM  ***   INCREASE M IF NECESSARY  ***
112 IF M>=B(J) THEN 130
120 LET M=B(J)
125 GOTO 145
130 IF M>=C(J) THEN 145
140 LET M=C(J)
145 NEXT J
147 REM ***  MAJOR LOOP ENDS HERE  ***
150 PRINT M;"IS THE LENGTH OF THE LONGEST MONOTONE SUBSEQUENCE"
999 END
```

6.4 SUMMARY

1. When constructing an algorithm, it is important to break down the solution to the problem into a number of routines. If a routine is to be used more than once in an algorithm with no more than a variation in the parameters passed to it, it should be written as a subroutine.

2. In order to call a subroutine in a flow chart, we draw the following box,

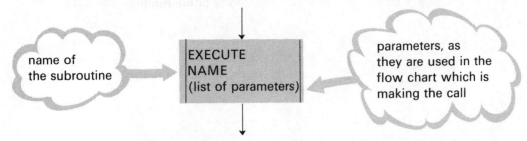

and, in written notation, we write the following.

execute NAME on (list of parameters)

In order to describe a subroutine by means of a flow chart, we draw the following boxes,

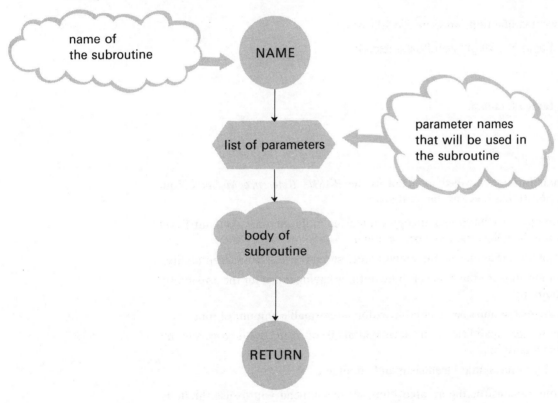

and, in written notation, we write the following.

begin NAME on (list of parameters)

.
.
.

body of subroutine

.
.
.

return

3. In order to describe a function routine by means of a flow chart, we draw the following boxes.

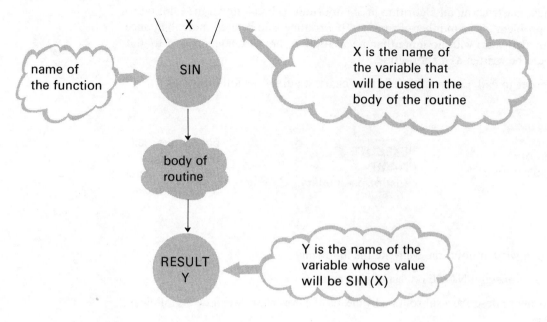

and, in written notation, we write the following.

> begin NAME on (list of parameters)
>
> .
> .
> .
>
> body of routine
>
> .
> .
> .
>
> result y

4. String handling is described in detail in the *BASIC Reference Manual*. You should refer to that book as the need arises.

5. In addition to the definition of an algorithm given in the summary of Unit 1, we also require the following to be true of them.

    (i)   An algorithm has one or more inputs which should be tested for acceptability.

    (ii) An algorithm produces one or more outputs having tested for the possibility of output.

    (iii) An algorithm must be executable within a reasonable amount of time.

    (iv) The storage capabilities of the device should be mirrored in the construction of the algorithm.

    (v)   An algorithm should be efficient and adaptable.

6. When you are constructing an algorithm, we recommend you to use the flow diagram which is printed at the back of this text.

Appendix I Proof of the Monotone Sequence Theorem

Since the longest monotone subsequence must be either increasing or decreasing, its length must be the largest of MAXINC and MAXDEC as computed by the algorithm of Figure 4-53, page A199. That is to say, the length of the longest monotone subsequence must be the largest of any of the B_J and C_K.

The whole crux of the proof then rests on the result of Question 2 of Exercises 4-5, Set A, page A198. To establish this result, suppose that $J < K$. Then, since $A_J \neq A_K$, we must have either

$$A_J > A_K$$

or

$$A_J < A_K.$$

If

$$A_J > A_K,$$

then

$$B_K \geq B_J + 1,$$

so that certainly

$$B_J \neq B_K.$$

If

$$A_J < A_K,$$

then

$$C_K \geq C_J + 1,$$

so that certainly

$$C_J \neq C_K.$$

We have, therefore, either

$$B_J \neq B_K$$

or

$$C_J \neq C_K$$

and hence (B_J, C_J) and (B_K, C_K) cannot be the same. We now ask how many different pairs (B_I, C_I) we can have.

Suppose the longest monotone subsequence to be of length M. Then we have

$$\text{all } B_I \leq M(I = 1(1)N)$$

and also

$$\text{all } C_I \leq M(I = 1(1)N).$$

There are thus only M possible values $(1, 2, \ldots, M)$ for each of the B_I and the C_I. So, the maximum number of distinct pairs (B_I, C_I) is M^2.

Since all N of the pairs

$$\{(B_I, C_I), I = 1(1)N\}$$

are distinct, we must have

$$M^2 \geq N$$

and hence

$$M \geq \sqrt{N},$$

as required.

Unit 7   Computer Hardware

Contents

Set Books

A. I. Forsythe, T. A. Keenan, E. I. Organick and W. Stenberg, *Computer Science: A First Course* (John Wiley, 1969).

A. I. Forsythe, T. A. Keenan, E. I. Organick and W. Stenberg, *Computer Science: BASIC Language Programming* (John Wiley, 1970).

It is essential to have these books to study this course. Throughout the correspondence texts, the set books are referred to as

> **A** for *Computer Science: A First Course,*
> **B** for *Computer Science: BASIC Language Programming.*

Notation

While we have used the notation of the set books for the representation of algorithms we differ from them in the use of the symbols

> O for the capital letter O,
> Ø for zero.

This has been done to bring our texts into line with the notation you will use when writing programs. In addition, we sometimes use the symbol

> ▽ for a space

in a string of symbols.

7.0 INTRODUCTION

One might think that to construct a BASIC program requires little knowledge of how a computer executes instructions and that what is important in writing a program is a thorough grasp of the allowable instructions. Thus what a particular computer actually does with the instructions we write may appear to be of mere passing interest. Nothing could be further from the truth. To gain a good appreciation of the usefulness and limitations of a given computer one needs a sound understanding of what processes it will carry out efficiently and how. In this unit we look at how a computer works, not as a piece of electronic equipment but as a logical machine that executes instructions. The physical components of a computer and the way they interact are called hardware to distinguish them from software, that being programs which use the hardware and, in particular, programs which enable us to make efficient use of a computer. In addition, it ought to be said that the philosophy behind the course, an *algorithmic* approach to computing, dictates that we are not even interested in the electrical devices and circuits which comprise merely one part of a computer's hardware. Rather our interest lies in describing the problem—How does a computer execute an instruction?—and in seeking the algorithm which solves it.

No two types of computer are the same in every detail, yet we felt that this unit required a description of a specific machine. We could have described a fictional machine, such as the SAMOS model described by **A**, but decided instead to look at a simplification of an actual machine. The reason for simplifying the descriptions we give is purely to save time. A programming course to teach all the various types of instruction that a general purpose computer can execute might take six weeks of full-time study, not the single week that you will spend on this unit working part-time. We hope that by concentrating on the essential features and ignoring many details we shall give you a clear understanding of the machine with the minimum of tedium. To reduce the bias that inevitably comes from looking at a specific machine, in section 7.4 we discuss some alternative types of computer.

7.1   THE ESSENTIAL FEATURES OF A COMPUTER

To identify the features without which a computer could not satisfactorily execute instructions, we need to go back to the beginning of the course. In *Unit 2, BASIC Programming I*, we saw that the BASIC language includes instructions of the following types.

> Assignment
>
> Decision
>
> Input and Output
>
> Branch

You know from practical experience that some computers can be used to execute a BASIC program, but the above list of instructions does not indicate exactly what is required of a computer beyond the fact that it must be provided with the means of interpreting instructions given in the form in which we write them.

When assignment was first discussed in **A**, it was explained in terms of a master computer with two slaves, the assigner and the reader. This conceptual model is useful as an aid to understanding assignment but loses its value if we try to use it to describe a computer, because in describing a computer in human terms we are likely to endow it with human characteristics, such as intelligence, free will and even memory, which would be wrong. To a computer, a program is merely a string of symbols. Whereas we can use our intelligence to interpret symbols, all that a computer can do is to manipulate them according to rules which we lay down.

How much can be deduced about computers from our limited experience of using them? The list of instructions used for writing a BASIC program includes input and output facilities. If we are to be able to use the computer to solve problems we must be able both to communicate data and instructions to it and to receive some results in return.

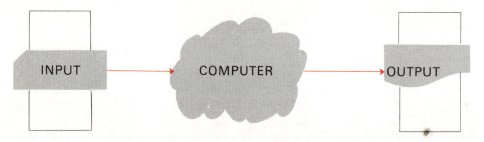

It is clear that, once we have input a string of symbols, the computer must be able to store them in some way or other. Otherwise, for instance, it would not be able to execute a loop of instructions. Even though it is not clear and, in fact, not necessary that a computer should store the actual symbols that we use, some representation of them must be recorded in a store and, equally, for the purposes of output it must be possible to retrieve information from the store and to display or print it appropriately.

Exercise 1

What can you say about the way a computer must be able to record and retrieve information and what are the implications on the organization of a computer's store imposed by the necessity of recording and retrieving information?

Solution 1

A computer can store both program instructions and data. Further, data can be of different types. For example, it may be a number, it may be a string of characters. Our experience tells us nothing about the way in which data is represented and stored within a computer. But we realize that the hardware must provide facilities for it.

An important point is that since data can be retrieved, each location, or sub-unit of the store, in which an item is stored must be identified by a unique name or address. Moreover since a computer can only handle an array of numbers or a string of characters these addresses must be in some sort of sequence.

As far as the program instructions are concerned we know that the computer can record them, and that the storage locations must be identifiable so that the instructions can be retrieved many times. Once again, this observation tells us nothing about the way instructions are represented and stored in the computer, simply that the hardware available must provide means of storing a representation of instructions in addressed locations.

We cannot deduce the sort of items that can be stored in each location, but we can learn several things from looking at the execution of an assignment such as

        4Ø LET A = B + C

if we assume that the value of each of the variables A, B and C is stored in a specified location.

We can see that assigning an item to a location destroys the previous contents of that location but retrieving an item does not change the contents of the location in which it is stored.

We can deduce that a computer is able to store the data and instructions that we give it, though we cannot deduce how these are stored. We also know that a computer can execute certain instructions; so a diagram like the following,

although representing some of the features of a computer, does not give the whole picture. What can we deduce about a computer from the execution of instructions?

Because you have had the experience of a computer executing BASIC instructions and know that computers are electronic devices, you might assume that a computer has special electronic circuits to execute BASIC instructions. But remember that a computer treats each BASIC statement as input data to another algorithm, the compiler, which breaks down each statement into several simple machine instructions. It is these instructions which the machine actually executes. Although we cannot deduce the exact form of these instructions we know that to execute them the computer must have an arithmetic unit which must be connected to the store. The arithmetic unit will be able to accept data from the store, process it and return the result to the store.

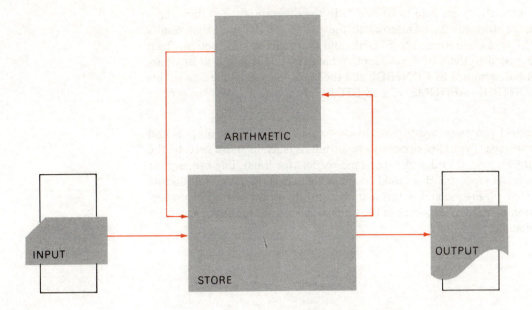

This is still not a complete picture of a computer. Consider the process of tracing an algorithm described in Unit 1, section 1.1.3. Tracing is the step-by-step execution of an algorithm by hand, but there is one activity involved in tracing a flow chart which is not trivial and this is the process of interpreting the instruction itself. To trace an algorithm we must be able to identify the type of each instruction and break it down into a sequence of simple steps. In a similar way a computer must have a control unit both to interpret instructions and select what data they use, and to keep track of which instruction to execute next. We can accommodate these facilities in a diagram as follows.

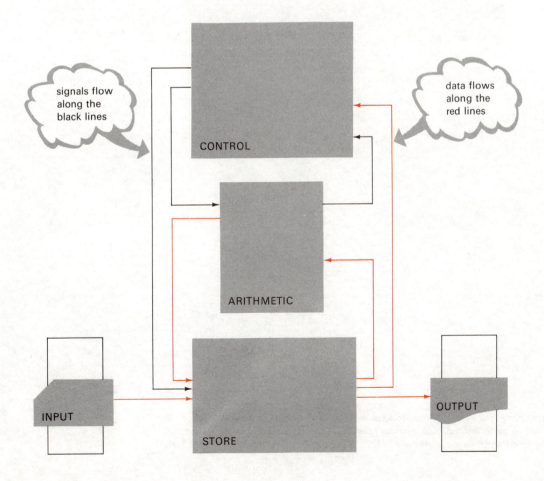

Both data and instructions are held in STORE*. In the diagram, the red lines show that data from the store can be transferred to the ARITHMETIC unit, that results from ARITHMETIC can be returned to STORE, and that machine instructions from STORE can be passed to the CONTROL unit. Which red line is in use at any particular moment is determined by CONTROL and the black lines show the flow of this information. CONTROL, ARITHMETIC and STORE are together called the central processor.

Outside the central processor are peripheral devices used for input, output, and supplementary storage. Typical peripherals are terminals (such as those used by the *Student Computing Service*), card and paper tape readers for input, line printers for output, and magnetic tapes and discs used to supplement STORE. We shall discuss peripherals in *Unit 10*, *Files I*, and in this text concentrate on the central processor. In the next section we shall discuss each of the units of the central processor and their links in more detail.

---

* Some words of warning. The field of computer science is full of ill-defined jargon. Throughout this course we have tried to use as few technical terms as possible and to be consistent in our use of them, but there are many terms which have one meaning in Britain and another meaning in the United States.

7.2 A TYPICAL COMPUTER

In this section we shall look at each of the essential components of the central processor in more detail. Rather than develop the SAMOS model described in **A**, which is rather unrealistic, we have chosen to describe in some detail the way in which some computers execute instructions. One such computer is the ICL 1903A, used by the Data Processing department of the Open University, and shown below.

Central
Processor

We do not expect this photograph to tell you very much. Certainly the cabinet of the central processor gives no indication of what it might contain. In this section we want to take the lid off and look inside. And yet it must be stated that what we shall have to say is a simplification of the real thing; but we have tried not to simplify matters to a state where they become unrealistic. We suggest that you read quickly through section 7.2. In section 7.3 we shall follow through the execution of some typical instructions. By seeing how the particular parts of the central processor interact, you should get a good feel for their individual purposes.

7.2.1 STORE

STORE in many modern computers uses magnetic cores and the one section of the SAMOS model that we recommend you to read describes a core store.

READ page A23, line 14, to page A31, line 2

Notes

Page A23 *line* −3   The word *memory* is used by **A** when we would use the word *store*. The description of a rectangular box should not be taken too literally.

Page A 24 *lines 3 and 6*   The SAMOS model has 61 core planes and 100 wires in each direction in a plane. These numbers vary from computer to computer.

As we see from the reading passage, information in a core store, and in most other forms of computer store, is represented by objects which can be in one of two states. If it were possible to manufacture a cheap, reliable object which was stable in more than two states, computer manufacturers would use them. Up until now, the only reliable objects are ones which can be in one of two states. We label the two states Ø and 1.

Thus data is held in a computer store as binary patterns, each digit of the pattern being called a bit. We need a code by which we can give a meaning to a binary pattern, thus turning data into information. We shall discuss briefly three codes in which a binary pattern stands either for a number, character or instruction. As described in **A**, the bits can be arranged in groups, each group being called a word. This arrangement allows for addressing the contents of store. Each word is identified by a unique address. In some machines, an address is given to each *character*—we shall discuss them in section 7.4; for the moment we restrict our attention to a store in which each *word* has an address. In the design of the computer that we shall describe each word has twenty-four bits, but this number varies from one computer to another; in the SAMOS model, a word consists of sixty-one bits (see page **A**29, Figure 1-24). Also, the number of words in its STORE varies from computer to computer. Indeed different 1903A's can have different numbers of words in STORE. The Open University 1903A (a medium sized data processing computer) holds 49,152 words in STORE.

We shall use the following type of diagram to represent part of the inside of STORE, and, for ease of reading only, we have divided each word into four groups of six bits.

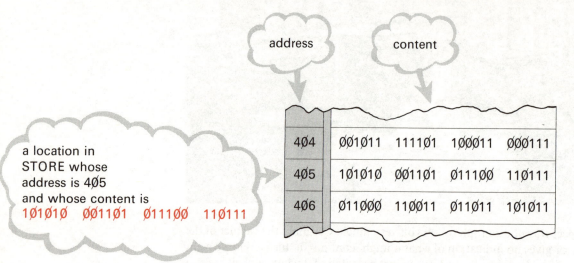

a location in STORE whose address is 4Ø5 and whose content is
1Ø1Ø1Ø ØØ11Ø1 Ø111ØØ 11Ø111

Several codes are used to represent information. The simplest of these is the number code, which uses each word to represent an integer by a binary pattern. Thus in the following diagram

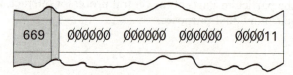

the number $+3$ is shown stored in the location whose address is 669; it is represented by the corresponding binary number*.

By convention, positive numbers use the last twenty-three bits; the first bit is always Ø. Negative integers are identified by the first bit being 1 and the rest of their representation is chosen to simplify the execution of arithmetic. Let us look at the representation of -3, which is

    111111 111111 111111 1111Ø1

If we consider the representation of $+3$ and -3 as being unsigned binary numbers and add them together, we have the following.

ØØØØØØ	ØØØØØØ	ØØØØØØ	ØØØØ11
111111	111111	111111	1111Ø1

1 | ØØØØØØ ØØØØØØ ØØØØØØ ØØØØØØ

* The binary number system is described in some detail in Refresher Booklet 12.

The result has twenty-four successive zeros and a leading 1 which is an overflow to the word-length of twenty-four. If this leading digit is ignored, the result of the addition is the representation of the number $\emptyset$. We have carried out the addition

$$3 + (-3) = \emptyset$$

For this reason, each negative number is represented by the complement* of the corresponding positive number with respect to 2^{24}. For, if the binary representations of a number and the negative of it are added together, the result is the representation of zero and an overflow.

This code also means that, in effect, the computer does not need to be able to subtract. By using complements, all subtractions can be turned into additions.

Using one word to store a number is a natural way of organizing STORE, but if we want to store a string of characters a new code is needed since, it is wasteful of space to put a single character in each word. With six bits, 64 different characters can be represented, sufficient for the essential characters of a typewriter keyboard (excluding lower-case letters). The following table gives the character code we shall use.

Six Bit Character Code

Code	Character	Code	Character
$\emptyset\emptyset\emptyset\emptyset\emptyset\emptyset$	$\emptyset$	$1\emptyset\emptyset\emptyset\emptyset\emptyset$	@
$\emptyset\emptyset\emptyset\emptyset\emptyset1$	1	$1\emptyset\emptyset\emptyset\emptyset1$	A
$\emptyset\emptyset\emptyset\emptyset1\emptyset$	2	$1\emptyset\emptyset\emptyset1\emptyset$	B
$\emptyset\emptyset\emptyset\emptyset11$	3	$1\emptyset\emptyset\emptyset11$	C
$\emptyset\emptyset\emptyset1\emptyset\emptyset$	4	$1\emptyset\emptyset1\emptyset\emptyset$	D
$\emptyset\emptyset\emptyset1\emptyset1$	5	$1\emptyset\emptyset1\emptyset1$	E
$\emptyset\emptyset\emptyset11\emptyset$	6	$1\emptyset\emptyset11\emptyset$	F
$\emptyset\emptyset\emptyset111$	7	$1\emptyset\emptyset111$	G
$\emptyset\emptyset1\emptyset\emptyset\emptyset$	8	$1\emptyset1\emptyset\emptyset\emptyset$	H
$\emptyset\emptyset1\emptyset\emptyset1$	9	$1\emptyset1\emptyset\emptyset1$	I
$\emptyset\emptyset1\emptyset1\emptyset$	:	$1\emptyset1\emptyset1\emptyset$	J
$\emptyset\emptyset1\emptyset11$	;	$1\emptyset1\emptyset11$	K
$\emptyset\emptyset11\emptyset\emptyset$	<	$1\emptyset11\emptyset\emptyset$	L
$\emptyset\emptyset11\emptyset1$	=	$1\emptyset11\emptyset1$	M
$\emptyset\emptyset111\emptyset$	>	$1\emptyset111\emptyset$	N
$\emptyset\emptyset1111$	?	$1\emptyset1111$	O
$\emptyset1\emptyset\emptyset\emptyset\emptyset$	Space	$11\emptyset\emptyset\emptyset\emptyset$	P
$\emptyset1\emptyset\emptyset\emptyset1$	!	$11\emptyset\emptyset\emptyset1$	Q
$\emptyset1\emptyset\emptyset1\emptyset$	"	$11\emptyset\emptyset1\emptyset$	R
$\emptyset1\emptyset\emptyset11$	#	$11\emptyset\emptyset11$	S
$\emptyset1\emptyset1\emptyset\emptyset$	£	$11\emptyset1\emptyset\emptyset$	T
$\emptyset1\emptyset1\emptyset1$	%	$11\emptyset1\emptyset1$	U
$\emptyset1\emptyset11\emptyset$	&	$11\emptyset11\emptyset$	V
$\emptyset1\emptyset111$	'	$11\emptyset111$	W
$\emptyset11\emptyset\emptyset\emptyset$	(	$111\emptyset\emptyset\emptyset$	X
$\emptyset11\emptyset\emptyset1$	)	$111\emptyset\emptyset1$	Y
$\emptyset11\emptyset1\emptyset$	*	$111\emptyset1\emptyset$	Z
$\emptyset11\emptyset11$	+	$111\emptyset11$	[
$\emptyset111\emptyset\emptyset$	,	$1111\emptyset\emptyset$	$
$\emptyset111\emptyset1$	–	$1111\emptyset1$	]
$\emptyset1111\emptyset$	.	$11111\emptyset$	↑
$\emptyset11111$	/	111111	←

* This is the number found by subtracting the corresponding positive number from 2^{24}. We leave you to convince yourself that the complement can be created by the simple rule "change each $\emptyset$ to 1, each 1 to $\emptyset$ and add 1 to the result".

As only six bits are used to represent each character, four characters can be represented in a single word. Thus in the following diagram

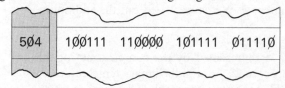

the characters

GPO.

are shown stored in the location whose address is 5Ø4. If the same binary pattern were to be interpreted as an integer, it would be $-6,353,954$.

Also, we need an instruction code since instructions too are held in STORE. We leave a full description of the code used for instructions until we have examined the nature of the instructions which our computer executes. For the moment, suffice it to say that the machine instructions we shall use have two parts, and will be stored as follows

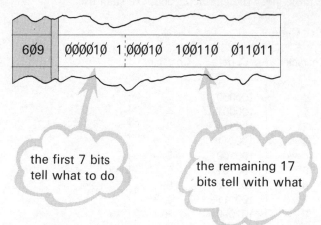

the first 7 bits tell what to do

the remaining 17 bits tell with what

Exercises 2–3

2. What is stored in the following words,

    (i) interpreting each word as a binary number,

    (ii) interpreting each word as four characters?

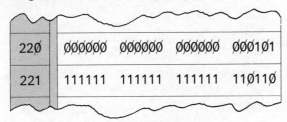

22Ø	ØØØØØØ	ØØØØØØ	ØØØØØØ	ØØØ1Ø1
221	111111	111111	111111	11Ø11Ø

3. How would you represent in store the message

    ANS ▽ + ▽143

    using the six bit character code?

Solutions 2–3

2. (i) Word 22Ø contains the binary number 5. The complement of the binary number in word 221 is ØØØØØØ ØØØØØØ ØØØØØØ ØØ1Ø1Ø which is the number +1Ø. Word 221 therefore represents −1Ø.

   (ii) Using the character code, word 22Ø represents Ø Ø Ø 5, word 221 represents ←←← V.

3.

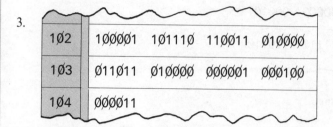

Any suitable addresses can be used. The last eighteen bits of the word 1Ø4 are not used in representing this string of characters.

Exercise 2 emphasizes the fact that the same binary pattern can be interpreted in different ways. The contents of a word never tells us what code to use to interpret it. Since the details of binary codes do not concern us in this course, we have acknowledged their existence but shall not use binary patterns in future. We shall however respect the forms of the different codes and will rewrite the binary pattern in three different ways.

(i) *Number Code*

If we are interpreting a word as a number, we shall write it in normal decimal form.

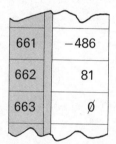

All you should note about this representation is that leading zeros are suppressed and only the negative sign is used explicitly.

(ii) *Character Code*

To show that we are implying the six bit character code, we divide the location in STORE with dashed lines into four sublocations.

Remember that although, in the above diagram, we show ANS ▽ in the location 1Ø2 the underlying binary pattern could equally well be interpreted by a different code to have a different meaning.

(iii) *Instruction Code*

Since, as we have said, an instruction has two parts, we shall divide the location in store with a single dashed line.

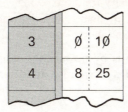

Each sub-location will contain a non-negative decimal number with leading zeros suppressed. The instruction held in location 4 above will be written 8 25.

So much for the way in which individual locations in store are used to hold information. In order to complete our picture of a store we need to look briefly at how STORE is used by the other components of the computer.

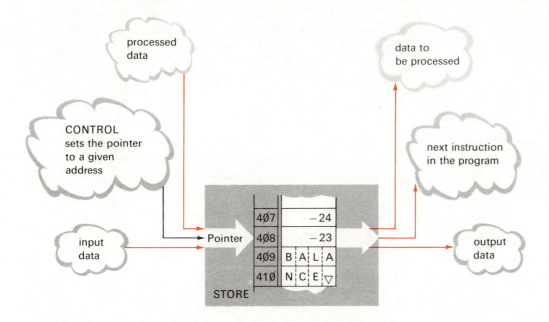

Only one word in STORE and only one path in or out of STORE can be used at a time. To change the address of the word in use CONTROL sends a signal for the Pointer to be moved. For instance the signal might specify address 4Ø8 and the Pointer would be moved to location 4Ø8. The next signal might be to record in location 4Ø8 a word of input data. The effect of recording this input data in location 4Ø8 would be to overwrite, and hence destroy, its previous contents. If, at a later stage in the program, we wanted to retrieve the data now stored in this word, the Pointer would have to be reset to location 4Ø8. The word could then be read without the contents of location 4Ø8 being destroyed.

Apart from the main store of the computer which we have been describing in this section, there are small temporary stores in the other units of the central processor. These small stores only need to be able to hold one word at any time and are called registers.

In conclusion, let us gather together the threads of this section. We have adopted the following conventions for describing STORE.

> STORE consists of addressed locations. In each location a word can be stored. A word consists of a binary pattern of twenty-four bits which may be the coded form of a number, a character or an instruction. A word can be recorded in or retrieved from the location indicated by a Pointer.

We hasten to add that this organization of a store is by no means common to all computers. It is however a realistic description of a currently* used machine. In section 7.4 we shall look at other possibilities.

---

* This unit went to press in 1972. The statement cannot be guaranteed true for all time.

7.2.2 CONTROL

How can the inanimate control unit issue directions and what form do they take? In principle, CONTROL is nothing other than an intricate collection of switches. Take the back off a modern computer you will find that STORE and ARITHMETIC are surprisingly compact but that the central processor cabinet is full of switching circuits. This is because CONTROL breaks down every machine instruction into a sequence of tiny steps to form what is called a microprogram. These steps are carried out by CONTROL opening and closing switches at the correct moments allowing data to flow.

The circuits in CONTROL serve to

(i)   interpret each instruction,

(ii) keep track of which instruction to execute next,

(iii) send signals to other parts of the computer indicating what operations to carry out.

We begin by looking at how instructions are interpreted. In *Unit 4, BASIC Programming II*, we saw that any expression, be it arithmetical, relational or logical, can be evaluated by means of a series of simple steps. Each simple step consists of evaluating an expression of the form

(operand) operator (operand)

or

operator (operand).

In fact we shall see later that, by making use of a special register in ARITHMETIC, called the Accumulator, all instructions can be reduced to the second form, namely

operator (operand)

or, in a loose sense,

do something with a specified something.

Such instructions are called single address instructions.

The first stage in the execution of a machine instruction is to copy the instruction from STORE to a register in CONTROL called the Instruction Register. The Instruction Register feeds a decoder which divides the instruction into two parts as follows.

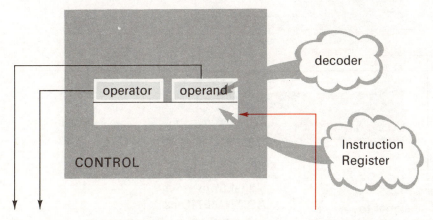

This is done automatically. An important point to remember is that in a machine instruction, the operand is an address. CONTROL uses the coded information in an instruction to record or retrieve the contents of some particular location in STORE. Thus a machine instruction might be

Output the contents of location 442.

One important feature of the central processor is that there must be a way of controlling the order in which instructions are executed. This can be done by using a special register in CONTROL called the Sequence Control Register (SCR).

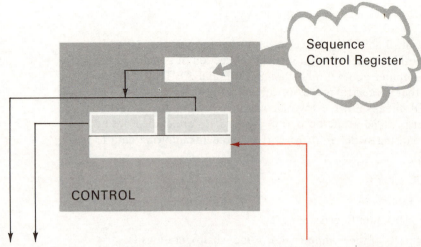

Sequence
Control Register

The content of the Sequence Control Register is always the address in STORE of the next instruction to be executed. The machine instructions are recorded in STORE in the sequence in which the program is to be executed. Every time an instruction is copied from STORE to the Instruction Register, the address in the Sequence Control Register is increased by 1 so that it contains the address of the next instruction in STORE.

To complete our picture of CONTROL, we need to allow for branching instructions. That is, we need a facility to reflect BASIC statements such as

        4Ø GOTO 1Ø
or
        4Ø IF X<Ø THEN 1Ø

To do the first of these, we need an instruction of the form

        Copy this address into the Sequence Control Register.

Thus we need a path connecting the operand to the Sequence Control Register. To do the second, we need to be able to record a signal which indicates whether a test failed or succeeded. This is done by an Indicator. The complete CONTROL can be described as follows.

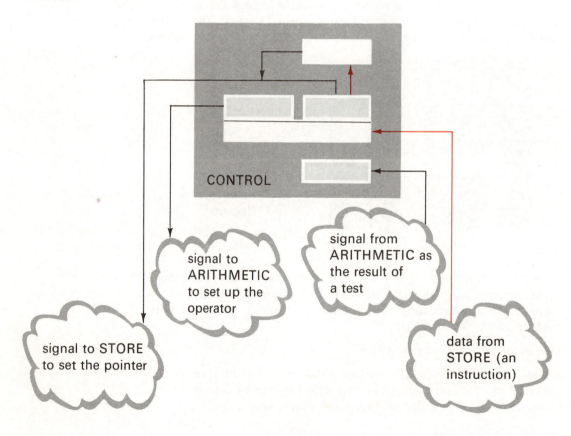

CONTROL

signal to STORE
to set the pointer

signal to
ARITHMETIC
to set up the
operator

signal from
ARITHMETIC as
the result of
a test

data from
STORE (an
instruction)

Exercise 4

Describe briefly how the components of CONTROL would be used for the execution of the following machine instruction.

> If the indicator has a specified value, the address of the next instruction in STORE is given by the address part of this instruction.

Solution 4

As a result of decoding the operator part of the instruction, CONTROL switches circuits causing the Indicator to be tested. If the test succeeds, the operand will be transferred into the Sequence Control Register; if it fails, this transfer will not take place. Thus, for example, the following diagrams give typical before and after situations.

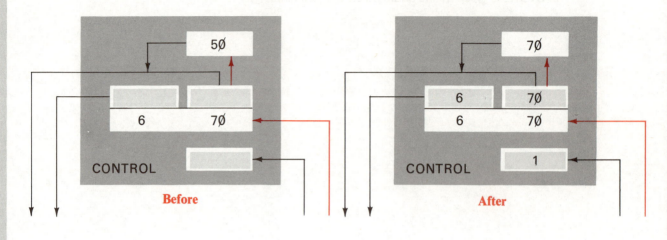

7.2.3 ARITHMETIC

As we saw in section 7.1, any computer must be able to process old data to give new. This is the purpose of ARITHMETIC and, from an intuitive point of view, it is in ARITHMETIC that the most obvious item of the computer's components is contained. This is called the Processor and should not be confused with the central processor of which it is a part.

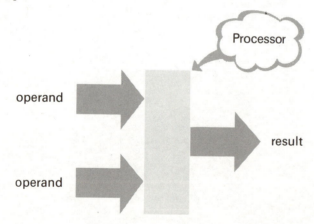

The Processor performs all the arithmetic operations that are included in our set of machine instructions, such as addition. We shall find that we have a large enough range of instructions to construct simple programs if the Processor can perform two binary and three unary operations. For the Processor to be able to perform a binary operation as shown by the above diagram it needs to be able to use two items of data at the same time. So part of ARITHMETIC consists of two registers for two operands.

These registers are called the Accumulator and the Buffer Register respectively.

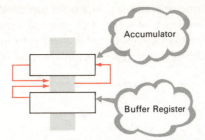

How the operands get into these registers is discussed in Exercise 5 at the end of this section. For the moment, let us say that the Buffer Register is always used for holding data which has just been transferred from store. The Accumulator is always used for holding the result of the operation just performed and it is from the Accumulator that data is transferred back into STORE. Follow the central three red lines and you will see how the execution of a binary operation is effected. The contents of both the Accumulator and the Buffer Register are fed into the Processor. The result of processing is fed back into the Accumulator.

To execute decision instructions we have to be able to transfer a signal from the Processor to CONTROL. This would be in response to a unary operation of the form

> Test the contents of the Accumulator and signal 1
> if zero and $\emptyset$ otherwise.

To do this, we add a signal line from the Processor to the Indicator into CONTROL. The complete ARITHMETIC can be depicted as follows.

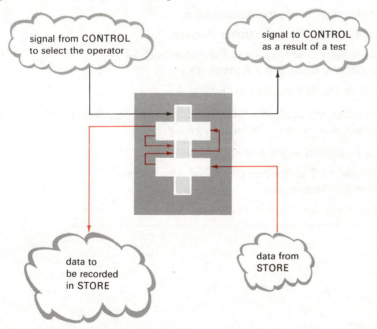

Exercise 5 (Hard—but read the solution)

How can ARITHMETIC be used to carry out an assignment such as*

using single address instructions only?

---

* In a flow chart the letters A, B and C refer to variables. In a computer the values of these variables are held in locations in STORE and in machine instructions we can only use the address of these locations. However instead of continually writing "the contents of the location in STORE where the value of the variable A is held", we simply write "the value of location A". Also, instead of writing "the address of the location in STORE which holds the value of the variable A" we write "A".

Solution 5

To use instructions with the address of only one operand (i.e., operator (operand)), we must break down the assignment into a sequence of steps. This is quite simply done by using an unusual instruction called the <u>load</u> instruction. Its effect is to load the contents of a specified address into the Accumulator. The <u>load</u> instruction is of the form

operator (operand)

since it is completely specified by

<u>load</u>   address

load into the Accumulator

the contents of a specified address

Once the Accumulator has been loaded, its contents can be used as the first operand in a two operand operation.

Since the assignment in the exercise refers to the addresses of three variables, at least three single address instructions will be needed. For the addition of B and C, the value of location B must be loaded into Accumulator and the value of location C must be copied into the Buffer Register. We could have the following four instructions.

1. Load the value of location B into the Accumulator.
2. Copy the value of location C into the Buffer Register.
3. Add the contents of the Buffer Register to the contents of the Accumulator, leaving the result in the Accumulator.
4. Store the contents of the Accumulator in location A.

The third of these instructions does not require a STORE address and in practice instructions 2 and 3 are combined into a single instruction giving

1. Load the value of location B into the Accumulator.
2. Add the value of location C to the contents of the Accumulator leaving the result in the Accumulator.
3. Store the contents of the Accumulator in location A.

7.3 INSTRUCTIONS

7.3.1 The Fetch–Execute Cycle

So far we have only described the individual components of the central processor. In this section we look at the way these components interact. This is best done by following how machine instructions are executed; that is, we shall discuss the operating cycle. As we have mentioned, the step-by-step execution of each machine instruction forms a microprogram, so it is our task to draw the flow chart of the microprogram which corresponds to a given instruction. In outline, the flow chart for each instruction is in two parts.

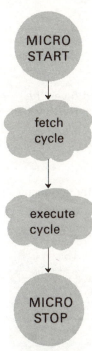

Since the machine instructions are held in STORE, the first of the two parts of the microprogram is to fetch the instruction from STORE and record it in the Instruction Register; this is called the fetch cycle. The second part is to execute the instruction; this is called the execute cycle. During the fetch cycle CONTROL sets two of the flow lines on the diagram.

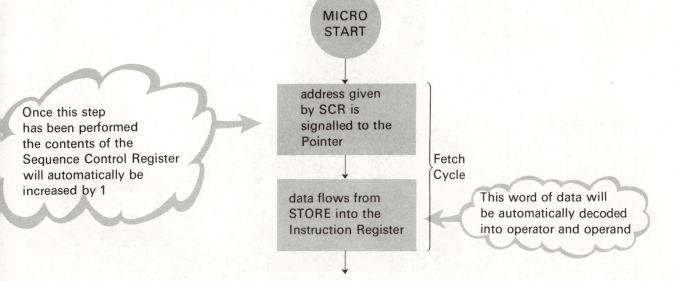

It is important to remember two things. At any instant, there is only one location in STORE to which the Pointer can be set. Thus the data which flows from STORE is the word to which the Pointer is currently set. Also the binary pattern of the operator code determines the subsequent action of CONTROL. No decision has to be made; the binary pattern simply sets a switch.

To expand the execute cycle, we clearly need to know what type of instruction is to be executed. Suppose that it is an input instruction. In coded form, it might be Ø 41, which we would represent in STORE as

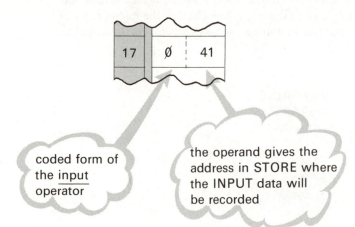

coded form of the input operator

the operand gives the address in STORE where the INPUT data will be recorded

At the end of the fetch cycle the computer might have the following values in its registers.

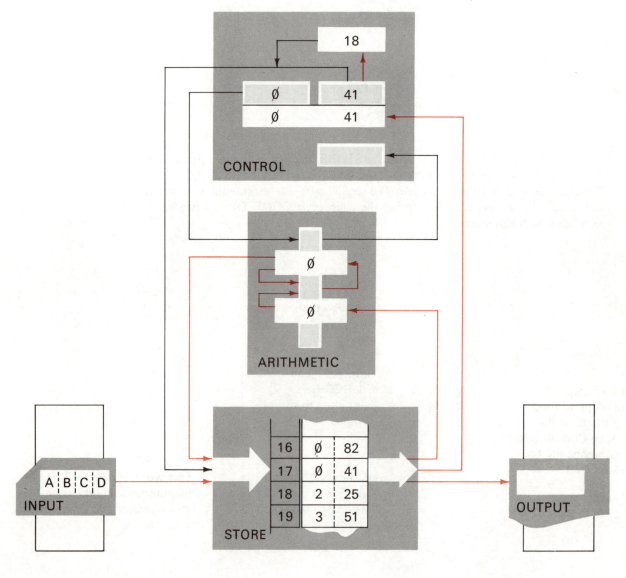

At the end of the fetch cycle we know that

(i)   the entry in the Instruction Register will be Ø 41.

(ii)   (Value of Sequence Control Register) = 1 + (Address to which Pointer is set).

The steps of the execute cycle are as follows.

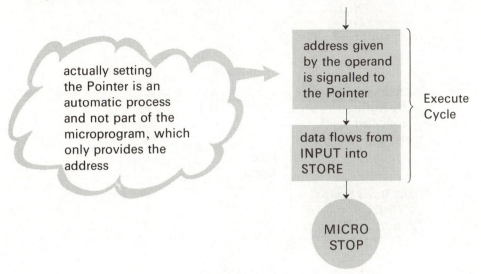

When these steps have been completed, we may have the following values in the registers.

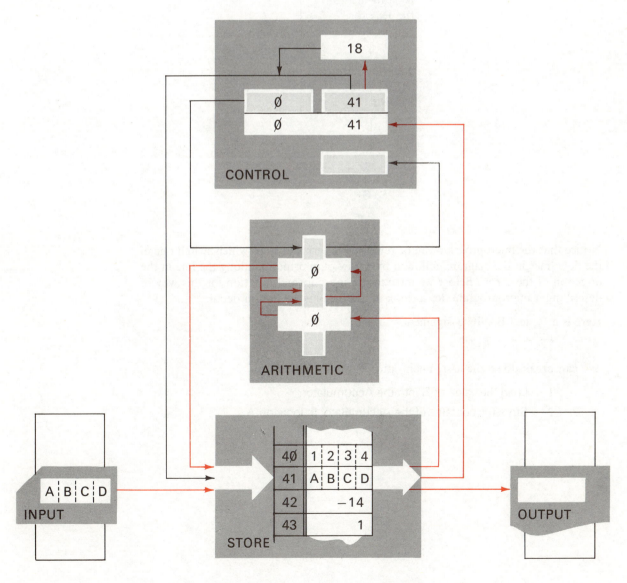

Of these values, only the ones in CONTROL and in the specified location in STORE are determined by the instruction. Here is the complete microprogram for an input instruction.

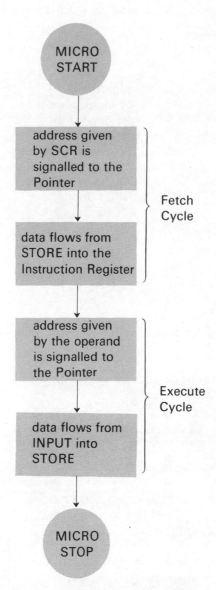

Notice that the microprogram can be rewritten in terms of steps which switch one of the flow lines in the diagram, followed by a flow of information along the line in the direction of the arrow. Before we introduce a more formal notation for this way of describing a microprogram, let us look at another instruction in detail.

Here is a typical BASIC assignment.

    1Ø LET A=B

We can break down the assignment into

      1.   Load the value of B into the Accumulator.

      2.   Store the contents of the Accumulator in location A.

Let us draw the flow chart of the microprogram for the first of these instructions.

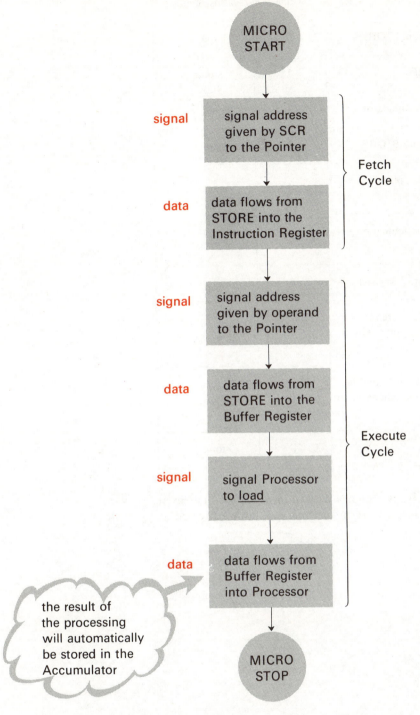

The microprogram for the second instruction is even simpler and we have left it for you to do as part of the next exercise.

Now that we have seen how an instruction is executed, we shall look at the complete set of machine instructions that we shall allow, and give the codes used for the operators. Remember that an instruction always consists of an operator followed by an operand. These instructions, although not precisely the same as the machine instructions of any particular computer, are typical of the instructions that many computers allow, except for input and output which have been deliberately simplified.

Interpretation	Operator Code	Written Form of Instruction
Read a word from INPUT and record it in STORE	Ø	input
Record in OUTPUT the word in STORE	1	output
Copy the content of STORE into the Accumulator	2	load
Copy the content of the Accumulator into STORE	3	store
Add the content of STORE to the content of the Accumulator	4	add
Subtract the content of STORE from the content of the Accumulator	5	sub
Store the operand in the Sequence Control Register	6	goto
If the content of the Accumulator is zero, store the operand in the Sequence Control Register, otherwise do nothing (branch if zero)	7	bz
If the content of the Accumulator is negative, store the operand in the Sequence Control Register, otherwise do nothing (branch if negative)	8	bn
Stop	9	stop

The form of each instruction in STORE is a binary pattern in which the first seven bits give the operator code and the last seventeen the operand, both expressed as binary numbers.

Exercise 6

Explain what the following words mean when interpreted as instructions in written form if the names A, B and C have been allocated to locations 54, 39 and 72 respectively.

(i)   2   39

(ii) 5   72

(iii) Ø   39

(iv) 6   25

(v)   8   54

Write down the 24-bit binary pattern used to represent each of these instructions and then interpret this pattern as four characters.

Solution 6

The written form of the instructions are as follows.

(i)   <u>load</u> B

(ii)   <u>sub</u> C

(iii) <u>input</u> B

(iv) <u>goto</u> 25

(v)   <u>bn</u> 54

These are represented by the following binary patterns.

(i)   ØØØØØ1   ØØØØØØ   ØØØØØØ   1ØØ111

(ii)   ØØØØ1Ø   1ØØØØØ   ØØØØØ1   ØØ1ØØØ

(iii) ØØØØØØ   ØØØØØØ   ØØØØØØ   1ØØ111

(iv) ØØØØ11   ØØØØØØ   ØØØØØØ   Ø11ØØ1

(v)   ØØ1ØØ   ØØØØØØ   ØØØØØØ   11Ø11Ø

Interpreting the above patterns as characters, we get the following.

(i)   1   Ø   Ø   G

(ii)   2   @   1   8

(iii) Ø   Ø   Ø   G

(iv) 3   Ø   Ø   )

(v)   4   Ø   Ø   V

Whereas we have been expressing microprograms by means of flow charts which show the flow lines set by CONTROL, there is a brief way of writing them down using the line numbers given on the diagram opposite. Each time a line number is written, it means that information at the start of the line flows to the end of the line and is recorded or acted upon appropriately. The following operations are carried out automatically.

(i)   The Sequence Control Register is increased by 1 each time a value is read from it.

(ii)   The Pointer is automatically set to the address signalled to it.

(iii) The result of processing any data in the Processor is stored in the Accumulator.

And, most important of all,

(iv) The binary pattern held in the Instruction Register is automatically* decoded into

        operator (operand).

---

* This means that CONTROL does not take decisions—its activities are predetermined by the binary pattern of the instruction that we input to STORE.

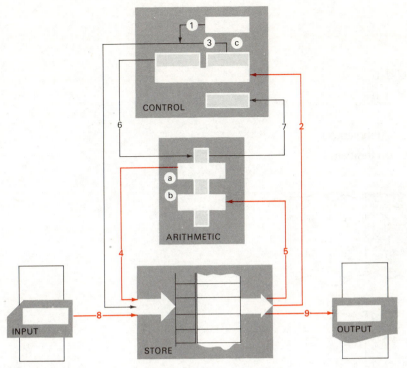

Thus the microprograms for the instructions that we looked at in detail on pages 28 and 29 are

> input   1, 2, 3, 8
>
> load   1, 2, 3, 5, 6, b

The first two digits are always 1, 2; they represent the fetch cycle.

Exercises 7–8

7.   Describe the microprograms for the other instructions that can be constructed from the allowable operators.

8.   Give the machine instructions which are equivalent to the following flow chart instructions. Express your answers in both written form and stored instruction code assuming that the value of variables A, B, C and D have been stored in locations 54, 39, 72 and 81 respectively.

    (i)   A ⟵ B

    (ii)   if A ≠ B then 5Ø

    (iii) A ⟵ B + C − D

    (iv)

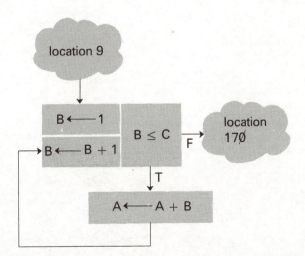

In each part the instructions are to be stored from location 1Ø onwards.

Solutions 7–8

7. <u>output</u>   1, 2, 3, 9
    <u>store</u>   1, 2, 3, 4
    <u>add</u>   1, 2, 3, 5, 6, a, b
    <u>sub</u>   1, 2, 3, 5, 6, a, b
    <u>goto</u>   1, 2, c
    <u>bz</u>   1, 2, 7, a, c (on Indicator)
    <u>bn</u>   1, 2, 7, a, c (on Indicator)
    <u>stop</u>

8. (i)   <u>load</u> B

1Ø	2	39
11	3	54

        <u>store</u> A

(ii) <u>load</u> A

1Ø	2	54
11	5	39
12	7	14
13	6	5Ø

       <u>sub</u> B
       <u>bz</u> 14
       <u>goto</u> 5Ø

(iii) <u>load</u> B

1Ø	2	39
11	4	72
12	5	81
13	3	54

       <u>add</u> C
       <u>sub</u> D
       <u>store</u> A

(iv) We need to have the number 1 in STORE. Suppose location 1∅∅ has the number 1.

B ⟵ 1	<u>load</u> 1∅∅	1∅	2	1∅∅
	<u>store</u> B	11	3	39
B ≤ C	<u>load</u> C	12	2	72
	<u>sub</u> B	13	5	39
	<u>bn</u> 17∅	14	8	17∅
A ⟵ A + B	<u>load</u> A	15	2	54
	<u>add</u> B	16	4	39
	<u>store</u> A	17	3	54
B ⟵ B + 1	<u>load</u> B	18	2	39
	<u>add</u> 1∅∅	19	4	1∅∅
	<u>store</u> B	2∅	3	39
	<u>goto</u> 12	21	6	12

7.3.2 Programming

Having described the model in some detail and having laid down the allowable types of instruction, we now turn to the problem of writing programs that the model can execute. However, before a program can be executed the instructions and constants that it uses must be held in STORE and locations must be set aside to hold the values of the variables in the program.

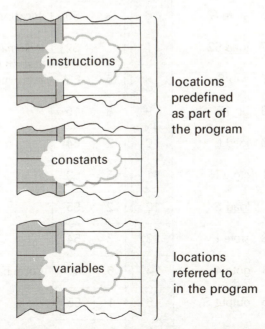

Instructions, constants and variables can be held anywhere in STORE so long as the instructions are in the correct sequence. Also, the binary patterns for instructions and constants have to be stored explicitly, whereas the initial values of variables do not. This means that no variable should be used until some value has been assigned to it by a store or an input instruction. If you have tried to use an undefined variable in a BASIC program, you will have seen a message of the type

UNDEFINED VALUE ACCESSED IN LINE 1Ø

In a program of machine instructions, you will not get an error message, but it is not sensible to use an undefined value since it could be anything—it will be the value left there by the previous program. But let us return to the matter in hand. In the following examples and exercises we use only the ten machine instructions defined in the previous section.

Example

Let us start by writing a program for one of the simple algorithms that you met earlier in the course—the algorithm to find the first Fibonnacci number greater than 1ØØØ (see page A16, Figure 1-10(b)). Below and opposite we have set out the algorithm

(i)   using the written notation described in Unit 1,

(ii)   using the written form of machine instructions,

(iii) using the coded form in which the program would be held in STORE.

Solution

Written Notation		Written Form		Coded Form			Comments
1.	N ← Ø	Ø	load 5Ø	Ø	2	5Ø	location 5Ø used for Ø
		1	store N	1	3	53	location 53 used for N
	L ← 1	2	load 51	2	2	51	location 51 used for 1
		3	store L	3	3	54	location 54 used for L
2.	S ← L + N	4	load L	4	2	54	
		5	add N	5	4	53	
		6	store S	6	3	55	location 55 used for S
3.	if S > 1ØØØ then 5	7	load 52	7	2	52	Alternative form of test must be constructed
		8	sub S	8	5	55	
		9	bn 15	9	8	15	
4.	N ← L	1Ø	load L	1Ø	2	54	
		11	store N	11	3	53	
	L ← S	12	load S	12	2	55	
		13	store L	13	3	54	
	goto 2	14	goto 4	14	6	4	Loop back
5.	output S	15	output S	15	1	55	
	end	16	stop	16	9		

50	0
51	1
52	1000
53	N
54	L
55	S

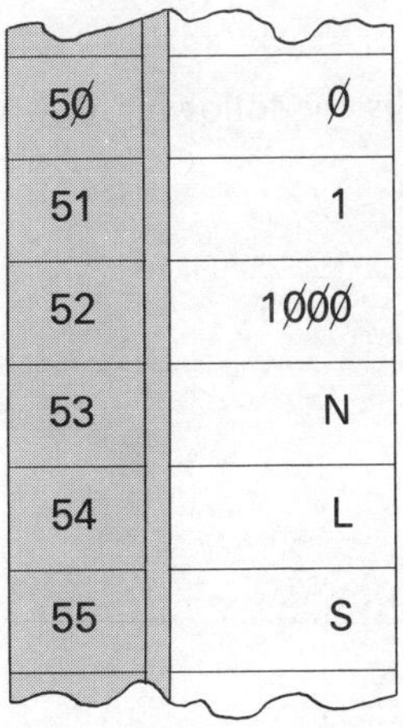

We have to store the constants and set aside locations for the variables which will be used in the program

---

Exercises 9–10

9. Look at the algorithm described on page A90, Figure 3-6. You may want to improve it slightly since the instructions in box 3 are repeated unnecessarily in boxes 7 and 8. Write the machine instructions for this algorithm using both the written form and the coded form.

10. Write the machine instructions for the Euclidean Algorithm (Subtraction form) given on page A101, Figure 3-11. (Try to avoid unnecessary goto instructions.)

Solutions 9–10

9. A better algorithm is described by the following flow chart.

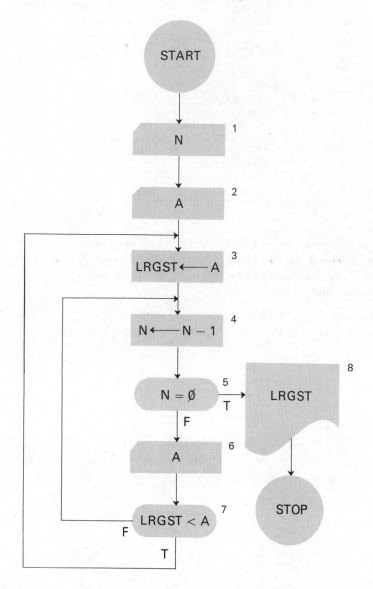

Here are the corresponding machine instructions.

Written Notation	Machine Instructions			
	Written Form	Coded Form		
begin				
1. input N	Ø input N	Ø	Ø	5Ø
2. input A	1 input A	1	Ø	51
3. L ⟵ A	2 load A	2	2	51
	3 store L	3	3	52
4. N ⟵ N − 1	4 load N	4	2	5Ø
	5 sub 53	5	5	53
	6 store N	6	3	5Ø
5. if N = Ø then goto 9	7 bz 13	7	7	13
6. input A	8 input A	8	Ø	51
7. if L < A then goto 3	9 load L	9	2	52
	1Ø sub A	1Ø	5	51
	11 bn 2	11	8	2
8. goto 4	12 goto 4	12	6	4
9. output L	13 output L	13	1	52
end	14 stop	14	9	

5Ø	N
51	A
52	L
53	1

10. We discussed the rearrangements of this algorithm which are necessary to avoid unnecessary GOTO instructions in Solution 12 of Unit 2, section 2.1.3.

Written Notation		Written Form		Coded Form		
begin						
1. input A, B	Ø	input A	Ø	Ø	:	52
	1	input B	1	Ø	:	53
2. L ← A	2	load A	2	2	:	52
	3	store L	3	3	:	5Ø
3. S ← B	4	load B	4	2	:	53
	5	store S	5	3	:	51
4. if L < S then goto 7	6	load L	6	2	:	5Ø
	7	sub S	7	5	:	51
	8	bn 13	8	8	:	13
5. L ← L − S	9	load L	9	2	:	5Ø
	1Ø	sub S	1Ø	5	:	51
	11	store L	11	3	:	5Ø
6. goto 4	12	goto 6	12	6	:	6
7. C ← L	13	load L	13	2	:	5Ø
	14	store C	14	3	:	54
8. L ← S	15	load S	15	2	:	51
	16	store L	16	3	:	5Ø
9. S ← C	17	load C	17	2	:	54
	18	store S	18	3	:	51
10. if S ≠ Ø then goto 5	19	load S	19	2	:	51
	2Ø	bz 22	2Ø	7	:	22
	21	goto 9	21	6	:	9
11. output L	22	output L	22	1	:	5Ø
end	23	stop	23	9	:	

Machine Instructions

All that we have done so far overlooks a basic problem. We have written out our programs as if they were stored in locations Ø, 1, 2, ... in STORE but we have not explained how the program gets into STORE in the first place. To understand how a program is put into store we must return to an outline of the fetch–execute cycle. We have added a START and a STOP point as shown below.

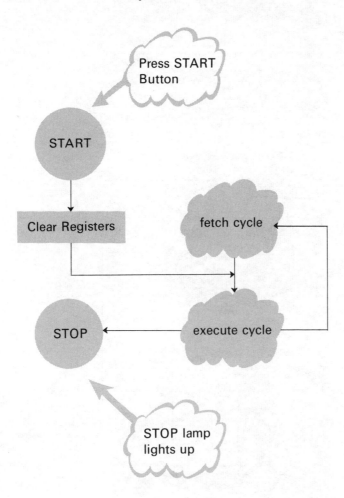

The start point shown in the flow chart corresponds to pressing a button to start the computer; the first step, Clear Registers, places zero in all the registers, and we then enter the normal execute cycle. The stopping point corresponds to the execution of a stop instruction in the program and for this reason it is shown emerging from the execute cycle. Its action is to light up a lamp to show that the computer has stopped. The following diagram shows what happens when the START button is pressed.

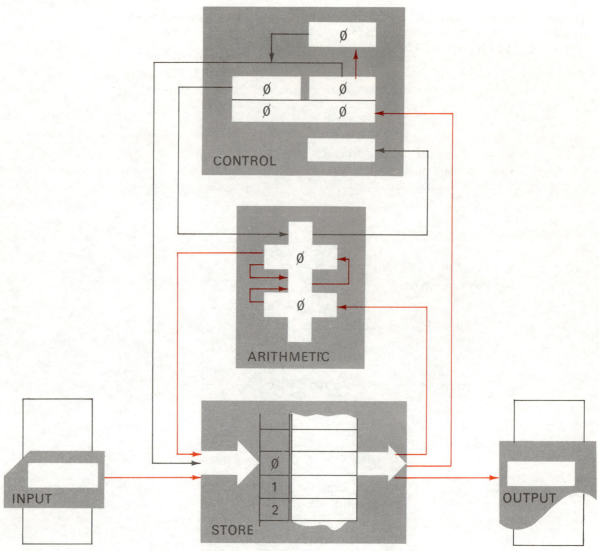

The first time we enter the execute cycle, we execute an instruction with operator and operand code Ø. The instruction being executed is, in fact,

> input Ø

This means that we must now provide an item of input data. Before deciding on the data to be provided let us consider what will happen next. We shall enter the fetch cycle for the first time, fetch the content of location Ø and decode it for execution as an instruction. So that the input data item must be an instruction. Suppose we input the instruction

> input 1

At this stage STORE will contain the following.

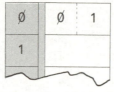

Thus returning to the first execute cycle, if we input the instruction

> input 1

the first fetch cycle will read this instruction and step the Sequence Control Register on to 1. We can now fill location 1 before its content is required as an instruction! Suppose we fill it with

> input 2

This will complete the second execute cycle and there will now be an instruction in location 1 waiting to be fetched.

We are certainly executing instructions, but not to any very useful purpose so far. What we want is a method of putting instructions elsewhere in STORE. This can be achieved as follows. Input to location 2 the instruction

goto Ø

For when this is executed the Pointer returns to location Ø. A complete cycle has now been set up. Can you see how to use it to input a program? Try the next exercise and study the solution carefully.

Exercise 11

The following instructions have been input to STORE.

Ø	Ø	1
1	Ø	2
2	6	Ø

When the instruction in location 2 is executed, the Sequence Control Register will be set to Ø. Thus the next instruction to be executed is in location Ø. This instruction is

input 1

Describe the input data you would present in order to read in a complete program and stop after the last instruction has been read in. Think of your solution in terms of the following steps.

(i)   What is the next item of input data?

(ii) What further data is required to read in and execute a complete program?

Solution 11

(i) We input to location 1 the instruction

input 5Ø

so that we can input a program instruction to location 5Ø when this instruction in location 1 is executed. Thus the input data is

Ø 5Ø

(ii) Following the execution of our latest instruction in location Ø, the situation in STORE is as follows.

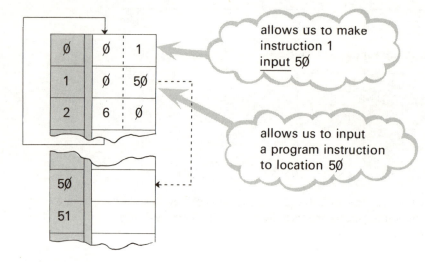

The next instruction to be executed is in location 1; thus the required data is as follows.

(first program instruction)

Ø 51

(second program instruction)

Ø 52

and so on

⋮

Ø 6Ø

(last program instruction)

6 5Ø

jump out of the input cycle to the first instruction of the program

The technique described in Solution 11 enables us to read a program and to store it in the computer. The first three instructions,

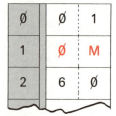

where M will be set to the required address each time the instruction

input 1

is executed, form a simple program whose effect is to load another program into

STORE. It is called a three address loader. In general, a program which is used to record the instructions of another program in STORE is called a loader. The three address loader is simple but it has drawbacks. For instance, the first four instructions will have to be repeated every time we write a program. Also, we have to provide an input instruction between each program instruction; if it were a long program, this would be a tedious thing to do. The alternative is to use a more complicated loader.

Also, if we do not want to input the loader before each program, we need an extra hardware facility, the RESTART button. The RESTART button causes execution to begin with a fetch cycle which fetches an instruction from the address given by the Sequence Control Register. This address can be set previously by the stop instruction, the full stop instruction being as follows.

STOP and store the operand in the Sequence Control Register.

In outline, the sequence of steps is given by the following flow chart.

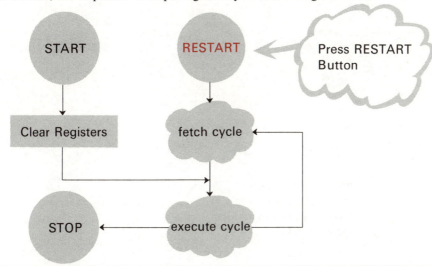

Exercise 12

The START button is pressed and the following input list is provided.

Ø	1
Ø	2
6	Ø
Ø	5Ø
Ø	56
Ø	51
2	56
Ø	52
8	55
Ø	53
1	56
Ø	54
6	5Ø
Ø	55
9	5Ø
9	5Ø
	9
	−2

(i) List the contents of STORE once the STOP lamp lights up.

(ii) What happens when the RESTART button is pressed?

The solution to this exercise is given on page 47.

Having input a program, we need some means of reversing the process, that is, we need some means to output the program. This process is best carried out by another program which is called a dumper. Thus we have the following programs in STORE.

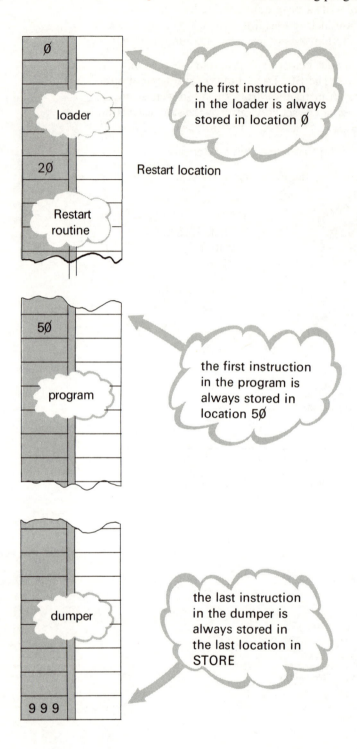

We have added to this diagram the position of the RESTART location. At the end of the loader, the program and the dumper we must include the instruction

stop 2Ø

Then, when the RESTART button is pressed, the instruction at location 2Ø will be fetched. If we wish to execute the program, then this should be done immediately, otherwise, we need to input

(i)   the address at which the loader or dumper should start,

(ii) the length of the program to be processed by the loader or dumper.

Solution 12

(i)

Ø	Ø	1
1	9	5Ø
2	6	Ø

5Ø	Ø	56
51	2	56
52	8	55
53	1	56
54	6	5Ø
55	9	5Ø

When the STOP lamp lights up, the instruction shown in location 1 above has just been executed. The value 5Ø has been stored in the Sequence Control Register, and the values 9 and −2 have not yet been used.

(ii) Execution of the small program stored in locations 5Ø to 55 above begins. The value 9 is read into STORE. Since it is not negative, 9 is output and the value −2 is read into STORE. This is negative and so the instruction

　　9　5Ø

in location 55 is executed. The STOP lamp lights up again.

In the next exercise, we ask you to write the machine instructions for the loader, the RESTART routine and the dumper. This is not a particularly easy exercise but one well worth taking some time over. If you find it difficult, consult the first part of the solution for part (i) and then try again.

Exercise 13

(i) Construct a loader.

(ii) Construct a RESTART routine.

(iii) Construct a dumper.

Hint: In the construction of these programs we have to use one binary pattern to mean two things. For example,

　　ØØØØØØ　ØØØØØØ　ØØ1Ø1Ø　ØØ11Ø1

means 653 when interpreted as a number and

　　input 653

when interpreted as an instruction.

Solution 13

(i) The following flow chart describes the main steps of the loader. This flow chart
is expanded into a complete set of machine instructions below

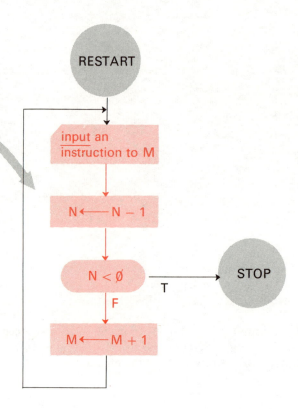

This loader will
not be executed
until M and N
have been assigned
values by the
RESTART routine

RESTART

input an
instruction to M

N ← N − 1

N < Ø

T → STOP

F

M ← M + 1

The Loader

Ø ⎫
1 ⎬ These locations have to be
2 ⎭ reserved for the three
address loader

3	input M
4	load 28
5	sub 13
6	bz 12
7	store 28
8	load 3
9	add 13
1Ø	store 3
11	goto 3
12	stop 2Ø
13	1

while N is in the
Accumulator, we might
as well use it!

store the number M in
location 3. When interpreted
as an instruction this
binary pattern is interpreted
as input M

(ii)

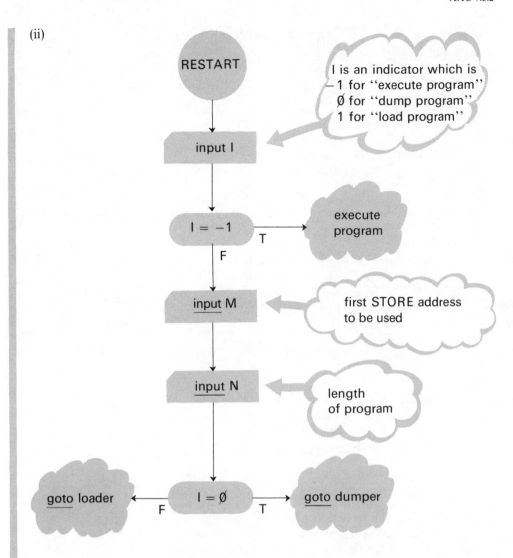

The machine instructions for the RESTART routine are as follows.

2∅   input 27
21   load 27
22   bn 5∅
23   input 3
24   input 28
25   bz 9
26   goto 3
27   I
28   N

(iii) The dumper is much the same as the loader. We must however bear in mind that
the operator code for <u>output</u> is 1.

The Dumper

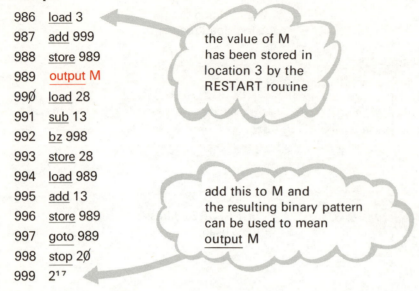

986	<u>load</u> 3
987	<u>add</u> 999
988	<u>store</u> 989
989	output M
990	<u>load</u> 28
991	<u>sub</u> 13
992	<u>bz</u> 998
993	<u>store</u> 28
994	<u>load</u> 989
995	<u>add</u> 13
996	<u>store</u> 989
997	<u>goto</u> 989
998	stop 20
999	2^{17}

the value of M
has been stored in
location 3 by the
RESTART routine

add this to M and
the resulting binary pattern
can be used to mean
<u>output</u> M

Both the loader and dumper operate on a consecutive sequence of locations in
STORE. Thus it is a good idea to use consecutive locations for the program instructions, followed by any constants that need defining. If in addition the variables are
allocated to the next locations in sequence the whole program (and nothing else)
can easily be dumped.

In this section we have met two important ideas. First, given an algorithm we can
systematically code the steps specified into machine instructions for our computer.
The translation process is straightforward, but laborious and tedious, and the resultant
program difficult to check. For this reason we would normally use a program (called
a compiler) to prepare programs for the machine by translating the corresponding
program written in BASIC or some other programming language into machine
instructions.

The second point is that in order to make use of a computer we must create special
programs, known collectively as software, which help us to use a computer effectively.
The combination of the computer hardware with the programs which make up the
software is referred to as a computer system. When you use a computer it is in fact the
system you use not simply the computer hardware.

7.4 OTHER TYPES OF COMPUTER

Sections 7.2 and 7.3 have concentrated on one design of computer but we do not want you to think that all computers are the same. They are not. Although the general principles of design are well established, the details differ considerably and in this section we shall look at the reasons for some of the differences which most affect a programmer.

The designers of computers choose a compromise between many factors. Cost is of great importance almost always, especially with small computers. Another consideration is what type of work a machine is designed to do. For instance, no scientific computer would be complete without provision for floating-point arithmetic but many small commercial computers work entirely in integers, or even digits. Commercial computers need to have high speed peripherals and a wide range of facilities for character handling. Scientific computers need fast arithmetic units.

7.4.1 Store

We know that computers handle and store two types of information: instructions and data on which the instructions operate. Although in theory there could be separate stores for instructions and data, both types of information are invariably held in a single store because of the flexibility it provides. This means that the formats chosen for representing instructions and data must be compatible and in particular the method of addressing should be suitable for both.

We have seen that, since computing is simply manipulating strings of symbols, STORE must therefore be able to hold such strings of symbols. There are two ways in which symbols can be recorded in identified locations and retrieved when required. The two ways differ only in the number of bits to which a given address refers; but once that difference has been made the knock-on effect to the representation of numbers, characters and instructions is considerable. A method of organizing STORE, which is called a character organized store, is to give the location in which each character is stored a unique address.

The variety of characters which can be represented depends on the number of bits provided. For instance with 6 bits we can form $2^6 (=64)$ distinct patterns and with 8 bits $2^8 (=256)$ different patterns. A typewriter keyboard usually has about 90 different characters including capital and small letters, digits and punctuation marks. When mathematical symbols and special characters to control the operation of peripherals are added even 7 bits (128 different characters) gives barely enough choice, so for a full character set 8 bits are needed. But for many purposes 6 bits are sufficient. In a character organized store each integer can be represented exactly using the smallest possible amount of store and instructions can be of various lengths. There are problems involved in performing arithmetic operations on integers stored in character form. The very large addresses met in a big store need many bits to represent them.

An alternative way of organizing the store is to group together a fixed number of consecutive storage elements into a single unit, called a word, and for each word to have an address. This is the system that we have discussed in the previous sections. The number of bits in each word is the same and is called the word length. Word lengths from 12 to 96 bits are in current use. In a word organized store, data and instructions can be represented in a straightforward manner if they fit within one word. In the organization we have discussed earlier each word can contain one of the following,

(i)   a 24 bit instruction,

(ii) an integer expressed as a binary number,

(iii) 4 characters.

More complicated items, such as floating point numbers, require several words. The advantages of word organization stem from the fact that data transfers between words and arithmetic operations on integers held in one word can be very efficient. Word organization is normal for small scientific computers, for process control computers and for very big machines in which fast arithmetic is all important.

Although electronic circuits can nowadays be made which are quite efficient at arithmetic on decimal numbers, the fastest computer arithmetic still operates on binary numbers. The largest binary number that can be stored in 6 bits is 63 and even in 8 bits only 255, so character organization is not suitable for binary arithmetic, unless several characters are grouped together to resemble a word. Since the input and output of numbers is most convenient if in decimal notation, binary arithmetic requires conversion of input numbers from decimal to binary and of results from binary to decimal before output. This conversion is usually quite slow but if long calculations are to be done its use may be justified by the extra speed of binary arithmetic. Since such calculations are common in scientific work but not in commercial work, scientific computers tend to have binary arithmetic whereas some commercial computers have only decimal arithmetic.

7.4.2 Instructions

There are certain types of instruction that computers execute. The type depends on the number of operands that may be associated with a given operator. Remembering that an operand is a means of specifying an address, an assignment such as

appears to require three addresses to be specified for the variables A, B and C. However, consider the following

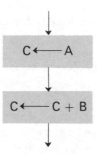

Each of these instructions requires only two addresses to be specified. They collectively express the same assignment, but need an extra instruction to do so. In fact we can break down these instructions into three steps.

> load A
>
> add B
>
> store C

Thus the assignment can be expressed in three different ways.

(i)   as a single instruction containing three addresses,

(ii) as two instructions each containing two addresses,

(iii) as three instructions each containing a single address.

Three address instructions are rarely used in computers, but single and two address instruction machines are common. Single address instructions are awkward with character organized machines but are a convenient form for a word organized machine. A character organized machine will normally allow for two address instructions.

In this brief section we have tried to show that, although all computers are merely processors of symbols, they do not all employ the same method of processing. This unit could equally well have described (and very nearly did) a character organized two address instruction machine. We felt however that the basic ideas common to all computers were more readily understood in relation to a word organized, single address instruction machine and accordingly adopted that point of view.

7.5 SUMMARY

1. The central processor of a computer consists of CONTROL, ARITHMETIC and STORE and is connected to the input and output peripherals.

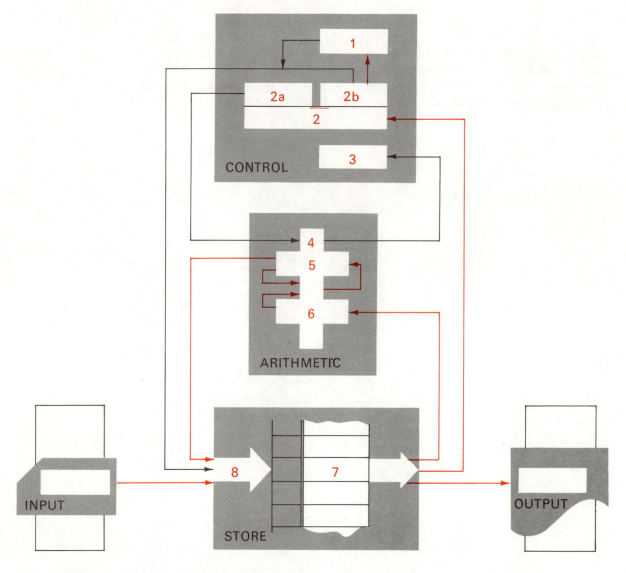

CONTROL contains the following items.

1. The Sequence Control Register, which holds the address of the next instruction to be executed.

2. The Instruction Register, which holds the current instruction that, in turn, is decoded into

> 2a. operator,
> 2b. operand.

3. The Indicator, which holds the signal from ARITHMETIC as to whether a test has succeeded or failed.

ARITHMETIC contains the following items.

4. The Processor which performs the arithmetic and logical operations.

5. The Accumulator, which holds the result of processing.

6. The Buffer Register, which holds the operands that have just been transferred from STORE.

STORE contains the following items.

7. Addressed locations, only one of which is current at any one time.

8. A Pointer which indicates the current location in STORE.

2. Information is held in STORE by objects which can be in one of two states, labelled $\emptyset$ and 1. Each state object is called a bit; 24* bits form a word. Each word is a binary pattern which can be coded as a number, 4 characters or an instruction.

3. Instructions are executed by means of a microprogram showing which of the flow lines are used to allow data (red lines) or signals (black lines) to flow round the diagram. The allowable instructions can be formed from the following operations.

Interpretation	Operator Code	Written Form of Instruction
Read a word from INPUT and record it in STORE	$\emptyset$	input
Record in OUTPUT the word in STORE	1	output
Copy the content of STORE into the Accumulator	2	load
Copy the content of the Accumulator into STORE	3	store
Add the content of STORE to the content of the Accumulator	4	add
Subtract the content of STORE from the content of the Accumulator	5	sub
Store the operand in the Sequence Control Register	6	goto
If the content of the Accumulator is zero, store the operand in the Sequence Control Register, otherwise do nothing (branch if zero)	7	bz
If the content of the Accumulator is negative, store the operand in the Sequence Control Register, otherwise do nothing (branch if negative)	8	bn
Store the operand in the Sequence Control Register and light up the STOP lamp	9	stop

* This number varies from computer to computer.

AN ALGORITHMIC APPROACH TO COMPUTING

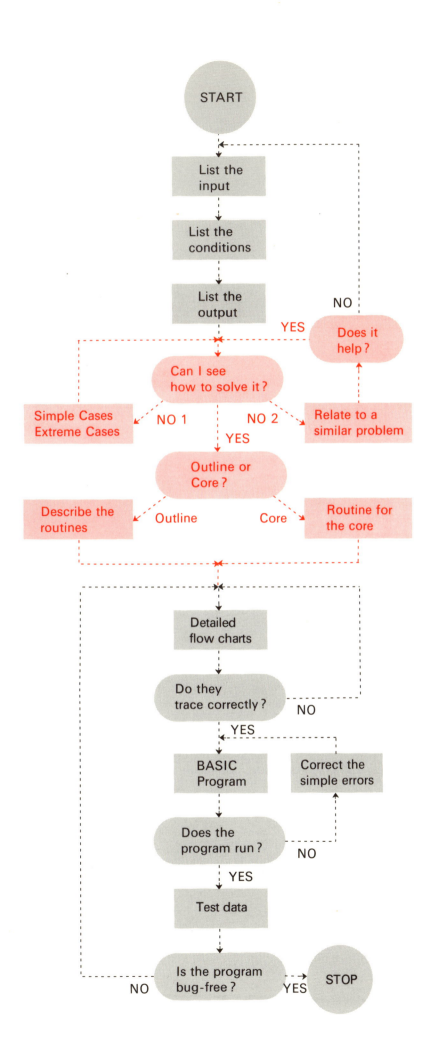